AF806816

Petroleum Engineering

Editor-in-Chief

Gbenga Oluyemi, Robert Gordon University, Aberdeen, Aberdeenshire, UK

Series Editors

Amirmasoud Kalantari-Dahaghi, Department of Petroleum Engineering, West Virginia University, Morgantown, WV, USA

Alireza Shahkarami, Department of Engineering, Saint Francis University, Loretto, PA, USA

Martin Fernø, Department of Physics and Technology, University of Bergen, Bergen, Norway

The Springer series in Petroleum Engineering promotes and expedites the dissemination of new research results and tutorial views in the field of exploration and production. The series contains monographs, lecture notes, and edited volumes. The subject focus is on upstream petroleum engineering, and coverage extends to all theoretical and applied aspects of the field. Material on traditional drilling and more modern methods such as fracking is of interest, as are topics including but not limited to:

- Exploration
- Formation evaluation (well logging)
- Drilling
- Economics
- Reservoir simulation
- Reservoir engineering
- Well engineering
- Artificial lift systems
- Facilities engineering

Contributions to the series can be made by submitting a proposal to the responsible publisher, Anthony Doyle at anthony.doyle@springer.com or the Academic Series Editor, Dr. Gbenga Oluyemi g.f.oluyemi@rgu.ac.uk.

Ehsan Khamehchi · Mohammad Reza Khaleghi ·
Amirhossein Abbasi · Javad Mahdavi Kalatehno

Applied Matrix Acidizing of Carbonate Reservoir

Springer

Ehsan Khamehchi
Department of Petroleum Engineering
Amirkabir University of Technology
(Tehran Polytechnic)
Tehran, Iran

Mohammad Reza Khaleghi
Department of Petroleum Engineering
Amirkabir University of Technology
(Tehran Polytechnic)
Tehran, Iran

Amirhossein Abbasi
Department of Petroleum Engineering
Amirkabir University of Technology
(Tehran Polytechnic)
Tehran, Iran

Javad Mahdavi Kalatehno
Department of Petroleum Engineering
Amirkabir University of Technology
(Tehran Polytechnic)
Tehran, Iran

ISSN 2366-2646 ISSN 2366-2654 (electronic)
Petroleum Engineering
ISBN 978-3-031-58280-6 ISBN 978-3-031-58281-3 (eBook)
https://doi.org/10.1007/978-3-031-58281-3

© The Editor(s) (if applicable) and The Author(s), under exclusive license to Springer Nature Switzerland AG 2024

This work is subject to copyright. All rights are solely and exclusively licensed by the Publisher, whether the whole or part of the material is concerned, specifically the rights of translation, reprinting, reuse of illustrations, recitation, broadcasting, reproduction on microfilms or in any other physical way, and transmission or information storage and retrieval, electronic adaptation, computer software, or by similar or dissimilar methodology now known or hereafter developed.
The use of general descriptive names, registered names, trademarks, service marks, etc. in this publication does not imply, even in the absence of a specific statement, that such names are exempt from the relevant protective laws and regulations and therefore free for general use.
The publisher, the authors and the editors are safe to assume that the advice and information in this book are believed to be true and accurate at the date of publication. Neither the publisher nor the authors or the editors give a warranty, expressed or implied, with respect to the material contained herein or for any errors or omissions that may have been made. The publisher remains neutral with regard to jurisdictional claims in published maps and institutional affiliations.

This Springer imprint is published by the registered company Springer Nature Switzerland AG
The registered company address is: Gewerbestrasse 11, 6330 Cham, Switzerland

If disposing of this product, please recycle the paper.

Letter of Commendation

With their profound and selfless embodiment of the concepts of sacrifice and self-sacrifice, with their boundless affection, and the comforting warmth of their presence, which serves as the strongest support during the coldest of times, with their compassionate hearts that weather tears and uncertainties, transforming fear into courage under their shelter, and with their unwavering, unconditional love that knows no bounds... I dedicate this collection to our beloved fathers and mothers.

A tribute to three sacred beings:
Those who grew weak so that we could gain strength...
Whose hair turned gray so that ours wouldn't...
And who burned with love to warm our existence and illuminate our path...

To our fathers,
To our mothers,
To our teachers.

Preface

Matrix acidizing is a widely used technique in the oil and gas industry to enhance well productivity and restore formation permeability. It involves injecting acid into the reservoir rock, which dissolves the carbonate minerals and creates wormholes that improve the flow of fluids. However, matrix acidizing is a complex process that involves unsteady, non-isothermal flow, chemical reactions, and changes in rock properties. Therefore, understanding the physics and chemistry of matrix acidizing is essential for optimizing its performance and profitability. This book is the result of our passion and curiosity for matrix acidizing in carbonate reservoirs. We have spent two years researching, writing, and compiling the most relevant and up-to-date information on this topic. Our goal is to provide a comprehensive and accessible reference for students, professionals, and researchers who are interested in learning more about matrix acidizing.

In Chap. 1, the journey begins with a fundamental exploration of matrix acidizing. The groundwork is laid by discussing formation damage, its types, and comprehensive considerations. This chapter navigates through the composition and characteristics of carbonate reservoirs and the intricate interactions between acids and rocks. The various types of acids used in acidizing and their effects on reservoir permeability and porosity are delved into. In Chap. 2, we venture into the realm of geomechanical modeling and its role in estimating formation fracture pressure. Elasticity parameters, uniaxial compressive strength, and pore pressure are explored. A comprehensive guide is provided for understanding the magnitude and direction of the main stresses in place. The intricacies of constructing a geomechanical model, considering factors like elastic constants, overburden stress, and the in situ stress regime, are introduced. Chapter 3 is dedicated to the rigorous examination of static and dynamic tests, essential for evaluating acidizing performance. Experiment design is discussed, and influential parameters like acid type, concentration, rock-to-acid ratio, and time are investigated. Through the lens of the rotary disk reactor, dissolution tests and the effects of rock porosity and mineralogy are explored. This chapter also discusses the application of various additives in acidizing and evaluates their compatibility and effectiveness. It culminates in guidelines for laboratory evaluation of acidizing additives, pivotal for success in acidizing operations. Chapter 4 takes us into the realm of software-driven analysis. The required information and the method of designing tests in software are discussed. The rational selection of experiment design levels and the considerations for acid injection rate and volume are thoroughly explored. This chapter scrutinizes the optimal acid selection and the insights gained from the software, bridging the gap between laboratory results and real-world application. In the final chapter, we delve into the economic considerations of acidizing. Understanding the essential costs associated with acidizing and the resulting flow rate changes is paramount. Through economic calculations, costs and profits resulting from acidizing operations are estimated. This chapter equips readers with the tools to make informed decisions, ensuring that acidizing is not only technically sound but also economically viable.

Tehran, Iran
Summer 2023

Ehsan Khamehchi
khamehchi@aut.ac.ir

Contents

Abbreviations

A	Cross Section
C	Adhesion
cm	Centimeters
F	Force
g	Gravitational Constant
in	Inches
L	Initial Length
L'	Final Length
MW	Molecular Mass
S_h	Horizontal Stress
UCS	Uniaxial Compressive Strength
V_p	Pressure Wave Velocity
V_s	Shear Wave Velocity
X	Volumetric Dissolution Power
β	Mass Dissolution Power
ΔL	Length Change
σ	Surface Tension
φ	Angle of Internal Friction

List of Figures

List of Tables

Chapter 1
Basic Objectives and Concepts of Matrix Acidizing

1.1 Introduction

Oil and gas reservoirs often face the problem of declining production over time due to various factors such as pressure drop, formation damage, permeability reduction due to pore plugging, etc. The low permeability of the reservoir rock is a natural production limitation, meaning that the fluid flow cannot move quickly from the reservoir to the wellbore. In addition, formation damage that occurs during operations, drilling, cementing, and production is another limitation to fluid flow from the reservoir to the wellbore. The methods used to solve these problems are called well-stimulation methods. Well-stimulation is used to reduce formation damage and increase well efficiency. Generally, different methods are used to stimulate the producing formation, such as matrix acidizing, hydraulic fracturing, and thermal stimulation. The choice of method depends on several factors such as the degree of damage, the depth of damage penetration, and the type of formation. Matrix acidizing is a common and well-known technique used in the oil and gas industry to improve reservoir productivity. This method involves injecting acid into the rock matrix at the well site to remove or reduce formation damage, such as drilling mud, sediment, and other solid materials that impede the flow of hydrocarbons (Ali & Ziauddin, 2020; Ali et al., 2020; Ali & Nasr-El-Din, 2020; Bazin, 2001; Beletskaya et al., 2017; Buijse & Glasbergen, 2005; Harry & Mcleod, 1984; McDuff et al., 2010; Panga et al., 2005; Taylor et al., 2006).

1.2 Formation Damage

Formation damage is a general term that refers to the disruption of the permeability of hydrocarbon formations caused by incompatible processes (Dargi et al., 2023; Kalatehno & Khamehchi, 2021). Formation damage is an operational and economic

© The Author(s), under exclusive license to Springer Nature Switzerland AG 2024
E. Khamehchi et al., *Applied Matrix Acidizing of Carbonate Reservoir*,
Petroleum Engineering, https://doi.org/10.1007/978-3-031-58281-3_1

problem that can occur during various stages of oil and gas production from underground reservoirs, such as exploration, drilling, hydraulic fracturing, and workover operations. This problem disrupts the permeability of the reservoir rocks, reducing the natural productivity of the reservoirs. It is therefore better to prevent damage than to try to repair it (Adair & Smith, 1994; Kalatehno & Khamehchi, 2021). The main objective of acidizing is to remove formation damage in the near wellbore. Therefore, it is important to understand the different mechanisms of formation damage in order to design acid formulations to remove the identified damage that reduces well productivity. It is important to note that not all damage can be removed by the use of acids, and only some types of damage can be removed by the use of acids. For example, the reduction in well productivity due to the reduction permeability caused by fine particles and clays can be removed by acid treatment, but the reduction in productivity caused by the reduction in relative hydrocarbon permeability may not be removed by acid treatment (Ali et al., 2016).

1.3 Types of Damage Formation Damage

Types of damage Formation damage can occur in two ways: natural and induced. Natural damage occurs as a result of production from the reservoir, while induced damage is caused by operations such as drilling, well completion, fluid injection, etc. It can be said that no operation does not have the potential to cause damage; therefore, when assessing formation damage, all aspects related to the well and its history should be examined and investigated (Xu et al., 2016; Zhao et al., 2019). Therefore, this section briefly describes some types of damage.

1.3.1 Natural Damage

1.3.1.1 Fine Migration

During production from the reservoir, grains and fine rock particles (like quartz or other similar materials) can be pushed out of the reservoir by the force of the overburden or the force of the production fluid and moved to places with smaller pores. This can cause fractures and pores to become blocked, causing pressure loss and damage to the formation. The fine migration particles also cause the formation of suspended particles in the production fluid, which accumulate as bridges in the pores of the wellbore and reduce well productivity (Byrne et al., 2014, Bassir et al., 2017).

1.3.1.2 Clay Swelling

Clay swelling is a type of formation damage where the permeability of the formation is reduced due to a change in the size of the clay grains. Clay swelling occurs when drilling, completion, workover, or stimulation fluids enter the formation and cause changes in the structure and chemical state of the clay. When exposed to water, clays swell due to their high water absorbency, reducing pore size and hence permeability, making fluid movement in the pores difficult and challenging (Azari & Leimkuhler, 1990a, 1990b; Chen et al., 2022; Kalatehno & Khamehchi, 2021).

1.3.1.3 Organic Deposits

One of the most common formation damage problems reported in oil-producing regions of the world is the formation of organic deposits in and near the wellbore. These organic deposits divide into two categories: paraffins and asphaltenes. The presence of heavy hydrocarbons such as paraffins and asphaltenes causes deposits to form in the tubing, near the wellbore, or in the formation as pressure and temperature decrease. The temperature drop can be caused by the injection of cold fluid near the wellbore, which will promote the formation of this type of deposit near the wellbore (Ali et al., 2016; Houchin & Hudson, 1986).

1.3.1.4 Mineral Deposition in the Formation Due to Water Injection

Formation damage in injection wells is a condition where the ability to inject into the well is reduced due to disruption of reservoir permeability in the wellbore or formation. Water injection is usually used for pressure maintenance, water disposal, or water flooding. Formation damage in injection wells can negatively affect the performance of water injection projects and increase project costs. Water injection into the reservoir can cause fine migration particles, the formation of insoluble mineral deposits, and the accumulation of solids in the oil. The formation of insoluble mineral deposits, which occur as a result of chemical reactions between the injection fluid and the formation water or rock, can include carbonates, sulfates, sulfides, iron compounds, and organic materials. These deposits can block pore throats, reduce effective porosity, and interfere with injection (Ali et al., 2016; Denney, 2013).

1.3.1.5 Emulsion

Emulsion refers to substances that are not soluble in each other in such a way that the droplets of one phase are dispersed and distributed in another phase, and this dispersion is not disturbed until the droplets coalesce or chemical substances (such as surfactant) are injected to reduce the surface tension of the droplets dispersed in the oil and cause the two phases to settle and separate. Emulsions present many challenges to oil and gas production and processing. They can increase fluid viscosity and density, reduce flow efficiency and pressure drop, interfere with oil–water separation, and reduce the injectability of injection wells. It is therefore necessary to prevent the formation of emulsions or to treat them effectively once they have formed. One of the most common methods of treating emulsions is the use of mutual solvents such as alcohols and surfactants (demulsifiers) to remove sediments from the wellbore area (Sousa et al., 2022).

Example 1.1 How does emulsion formation impact oil and gas production?

Answer:

Emulsions can increase fluid viscosity and density, reduce flow efficiency, interfere with oil–water separation, and decrease injection well injectability, resulting in production challenges.

1.3.1.6 Gas Condensate Pore Plugging

Formation damage due to pore plugging by condensate is a condition where gas production is reduced due to the accumulation of liquid hydrocarbons in the near wellbore or formation. Condensate blockage occurs in gas reservoirs or gas condensate reservoirs. In gas condensate reservoirs, when the reservoir pressure falls below the dew point, gas condensate is formed, which reduces the relative permeability of the gas and causes problems in gas production; this factor is also referred to as the phase-dependent skin effect. Furthermore, it causes liquid loading in such wells (Hassan et al., 2019, 2020).

Example 1.2 What are the main types of natural formation damage?

Answer:

Natural formation damage includes fine migration, clay swelling, organic deposits, and mineral deposition due to water injection, emulsion, and gas condensate pore plugging.

1.3.2 Induced Damage

1.3.2.1 Damage Caused by Various Operations

- **Deposits caused by acid treatment**

During the reaction of the acid with the formation rock, the acid may produce products that lead to the formation of sediments; therefore, the acid used should not produce insoluble sediments after reacting with the formation rock. In addition to the formation of sediments, the acid may alter the wettability of the reservoir rock, resulting in formation damage. Also, when acid is injected from the well into the reservoir, it can react with the tubing and produce iron deposits, which reduce the performance of the acid (Bryant & Buller, 1990; Radwan et al., 2022).

- **Damage Caused by Drilling**

The drilling mud used to drill a well must have properties such as viscosity, appropriate weight, and the ability to transport cuttings. For each of these factors, additives are added to the drilling mud, and each of these additives can cause damage to the well and the formation. In the drilling process, the invasion of solid mud particles into the formation can cause pore plugging and a reduction in the permeability. The invasion of drilling mud filtrate can also change the water saturation of the formation in the areas around the wellbore, which can reduce the relative permeability of the oil and consequently reduce production. At the same time, mud filtrate can also alter the wettability of the rock around the wellbore, causing clay swelling and pore blockage (Kalhor Mohammadi et al., 2023).

- **Damage from Completion and Workover Operations**

Completion and workover fluids are fluids used in various operations to increase well productivity. The main causes of formation damage caused by completion and workover fluids are pore plugging by solids, fluid/fluid or rock/fluid incompatibility, and wettability changes. One of the most common types of completion and workover fluids is water-based fluids, usually consisting of transparent brine. Often, transparent brine contains some solids, such as corrosion products, bacteria, and debris. Penetrating these solids into the formation causes formation damage. Therefore, it is important to control the amount of fine solids to minimize formation damage. Formation damage can reduce the natural productivity of reservoirs and increase well-stimulation and production costs (Hemeida & Gawish, 2008).

- **Damage Caused by Cementing**

Cementing is a critical operation in the drilling and completion of oil and gas wells. Cementing is used to seal the annulus between the casing and the formation, provide zonal isolation, support the casing, and protect the casing from corrosion. However, cementing can also cause damage to the formation and wellbore, affecting the productivity and integrity of the well. One of the main causes of cementing damage is the ingress of cement slurry into the formation. Cement slurry is a mixture of cement, water, and additives that is pumped into the annulus at high pressure. If the formation has high permeability or low fracture pressure, the cement slurry can penetrate the pores or fractures of the rock, reducing its effective permeability and porosity. This can affect the flow of oil and gas from the reservoir to the wellbore, resulting in lower production rates. Another cause of cementing damage is the corrosion of cement and casing by acidic formation fluids. Some formations contain corrosive fluids such as carbon dioxide, hydrogen sulfide, or organic acids that can react with the cement and casing, causing deterioration and weakening of their mechanical properties. This can lead to cracks, leaks, or failures in the cement sheath or casing, compromising the zonal isolation and structural integrity of the well. Corrosion can also create pathways for fluid migration between zones or to the surface, posing environmental and safety risks (Ali et al., 2016; Yan et al., 2019, 2020).

- **Damage Caused by Perforating**

Perforating is the process of making holes in the casing, cement, and formation to create pathways between the well and the reservoir. Perforating operations are performed in three ways: underbalanced, overbalanced, and a combination of the two. In an overbalanced condition, perforating at a pressure higher than the fracture pressure of the formation can result in the fracturing of the rock around the perforation and the intrusion of dirty fluids into the formation. If the density of perforations in the reservoir layer is less than or more than optimal, this will change the flow pattern around the wellbore and result in an additional pressure drop around the wellbore. Perforation is critical to well productivity, but the production of fine particles, damage to the cement, and the creation of a crushed or compacted zone around the perforation tunnel can damage the formation. Fines are small pieces of rock created by the high-pressure shock waves of perforating guns. Fines can block the perforation tunnel, migrate into the formation, and reduce permeability (Yan et al., 2019).

1.3.2.2 Wettability Alteration around the Wellbore Due to Drilling and Completion Fluids

Wettability is a property that describes how a fluid spreads and adheres to a solid surface in the presence of other immiscible fluids. The wettability of the reservoir rock is critical for different fluids and influences how the fluids are distributed in the pore space. This is because the wetting fluid tends to occupy small pores due to gravity. In addition, wettability affects properties such as relative permeability and capillary pressure. Therefore, proper knowledge of the wettability type of a reservoir is very effective for its optimal and proper exploitation (Dandekar, 2013; Morrow, 1998). Wettability alteration-induced formation damage is a condition in which the ability of the reservoir rock to produce hydrocarbons is reduced due to changes in the surface behavior of the rock. During the drilling and completion of a well, drilling fluid can enter the formation and change the wettability around the wellbore from water-wet to oil-wet, or vice versa. To prevent formation damage caused by wettability changes, it is possible to use fluids that are compatible with the formation properties and effective stimulation and enhancement methods that ensure uniform fluid distribution and contact time (Ballard & Dawe, 1988).

1.3.2.3 Bacteria

- **Aerobic Bacteria**

These are bacteria that require oxygen to grow. An example of aerobic bacteria is iron-oxidizing bacteria, which convert iron from the ferrous state to the ferric state. Ferric iron precipitates at a pH of about 2.2, whereas ferrous iron precipitates at a pH of about 6. Since the pH of spent acid rarely reaches 6, the ferric scale is rarely a problem, and the ferrous scale is the main cause of problems in acid operations.

- **Anaerobic Bacteria**

These are bacteria that do not grow in the presence of oxygen. Of the anaerobic bacteria, sulfate-reducing bacteria are the most problematic in the reservoir. As their name suggests, these bacteria reduce sulfate and sulfite in the water to sulfide and produce hydrogen sulfide, which increases the corrosiveness of the water and causes steel blister and cracking of the sulfide.

- **Facultative Bacteria**

These are bacteria that can grow in the presence or absence of oxygen. Sludge-forming bacteria are facultative bacteria. These sludges have a high density and cover the surface.

1.3.2.4 Pores Blocked by Water

Water can block pores in low-permeability rocks, preventing the flow and movement of hydrocarbons (This is due to the low mobility ratio and the large difference in viscosity, which prevent the displacement of oil by water.). When a large volume of drilling fluid or water is injected into the formation in the wellbore, a zone of high water saturation forms near the wellbore. In this zone, the relative permeability of the hydrocarbon phases decreases, resulting in a reduction or loss of well productivity (Abrams & Vinegar, 1985; Tannich, 1975).

1.3.2.5 Iron Deposition

When the acid is injected through the pipes into the well and reservoir, it will be in contact with these pipes, and the corrosive nature of the acid will cause a reaction with the pipes at the wellhead and in the well. The reservoir rock also contains iron and sulfur compounds, which increase the probability of iron deposits forming in the formation (Assem et al., 2019).

1.4 Formation Damage Assessment Comprehensive Considerations

Formation damage assessment is a critical step in determining the feasibility of acidizing well-stimulation operations. As a result of mechanical or chemical processes that cause solid particles to obstruct pores, formation damage reduces a reservoir's permeability. These particles can originate from external sources, such as the ingress of completion and workover fluids into the wellbore, or from internal sources, such as the movement of fines. It can be said that no operation is free from the risk of causing damage; therefore, all aspects related to the well and its history should be considered when assessing formation damage. The factors that should be considered when investigating formation damage are described in this section.

1.4.1 Reservoir Geology and Mineralogy

In the field of formation damage assessment, it is essential to obtain information on the geology and mineralogy of the reservoir. It is very important to know the type of rock (sandstone or carbonate), its permeability, and its porosity. This includes the distribution of porosity and permeability in the rock. In addition to the mineralogical characteristics of the rock and the amount and type of different minerals, it is very important to know the type, amount, and distribution of different rock minerals in the reservoir and the minerals around the pore or network of natural fractures. For example, in a carbonate rock, it is assumed that there is 95% calcium carbonate and 5% silicate minerals, but the distribution of these silicate minerals is such that they are only found within the walls of pores and natural fractures. When acid is injected (designed for a rock containing 95% carbonate minerals), the acid will first enter the pores and fractures and encounter non-carbonate minerals. This problem can lead to weak stimulation or detachment of these particles and their transport in the acid, causing pore plugging and reducing rather than increasing the permeability of the natural fractures. For this reason, an accurate description and understanding of the reservoir geology, including rock type, permeability, and porosity, as well as an accurate assessment of the mineralogical composition and distribution within the reservoir matrix and surrounding pore and fracture networks, are essential to a complete formation damage assessment.

Example 1.3 What information about reservoir geology and mineralogy is critical for formation damage assessment?

Answer:

Knowledge of the type of rock, its permeability, and porosity, including the distribution of these properties, as well as the type, amount, and distribution of different rock minerals within the reservoir and around pore networks or natural fractures, is critical.

1.4.2 Reservoir Fluids

Another factor that must be known in order to assess formation damage is knowledge of the reservoir fluids, which includes the fluid types (oil, gas, and water) and their characteristics. Fluid characteristics include the amount of CO_2 or H_2S in the gas, the specific gravity of the oil, the paraffin and asphaltene content of the oil, the amount of produced water and its characteristics, ionic compounds, and its tendency to scale. The analysis of reservoir fluid samples should also be considered. Reservoir fluids can cause damage due to the deposition of wax and asphalt from oil or precipitation from produced salt water. In addition, the evaluation of well characteristics and conditions should identify fluids that are incompatible with the reservoir fluids and should be considered in the design of operations. For this reason, a full assessment of formation damage requires a full understanding of reservoir fluid characteristics, including fluid types, associated properties, and possible damage mechanisms.

1.4.3 Production of Nearby Wells

In the initial phase of formation damage assessment, a comparative analysis of similar wells, including neighboring wells in the same field, provides valuable information on the suitability of the target well for stimulation. By comparing the production performance of these wells, indications of possible formation damage can be identified. In particular, a well that is producing at a lower rate than neighboring wells, despite favorable reservoir pressure, is a positive indicator of possible formation damage that could impede hydrocarbon flow and potentially reduce production. Therefore, evaluation of the production of neighboring wells, together with considerations of reservoir pressure differential and production disparities, serves as a valuable initial screening tool to identify potential formation damage and introduce candidate wells for subsequent stimulation.

1.4.4 Well Production History

A detailed examination of the well production history is of great importance. A sudden and rapid decline in production may indicate formation damage, usually caused by the movement of particles within the reservoir and their entrapment in pores. The production damage that can occur depends on the type of well (oil or gas) and the properties of the formation. Some types of damage include particle invasion, inorganic precipitation, and organic precipitation (such as paraffin and asphaltenes). Therefore, it is essential to perform a detailed analysis of the well production history to identify any significant changes or declines that may indicate the occurrence of formation damage. This investigation will enable the type of damage to be determined and facilitate the design of appropriate mitigation strategies tailored to the well's characteristics and operational requirements. Therefore, a comprehensive evaluation of the well's production history, with particular emphasis on sudden declines in production rates and types of formation damage, provides important information about the nature and extent of potential damage.

1.4.5 Drilling History

By studying the drilling history, it is possible, in some cases, to quickly identify the sources and causes of formation damage during drilling. This problem includes the identification of the fluids used and their characteristics (oil-based, water-based, pH, and solids used). In this section, two.

Mechanisms of formation damage are worth mentioning:

- Lost of drilling mud into the formation
- Invasion of drilling solids

Therefore, a detailed analysis of the drilling history, focusing on the above-mentioned mechanisms of formation damage, allows the rapid identification of potential sources and possible causes of formation damage. So, a full evaluation of the drilling history, including a study of the drilling fluids used and a mapping of the formation damage mechanisms, is important for figuring out what kind of damage may have been done to the reservoir and then fixing it.

1.4.6 Cementing Program

During the cementing operation, the probability of damage to the formation is lower than during other operations. However, the cementing program and its reports should be reviewed for follow-up and investigation of the following:

- **Excessive Amount of Cement with High pH**

The presence of an excessive amount of cement with high pH, especially in sandstone formations, can cause decomposition and the invasion of particles into the formation. Therefore, it is necessary to check the volume of cement and its effect on the reservoir integrity.

- **Invasion of Solid Materials and Filling of Pores**

Any invasion of solid particles of cement slurry into the formation or filling of pores requires a thorough investigation. This phenomenon can block fluid flow paths and put at risk the overall productivity of the well. Therefore, it is very important to evaluate the presence and distribution of cement-related solids in the reservoir matrix.

- **Cement Leakage**

The occurrence of complete leakage of cement to the bottom of the well or its invasion into natural fractures requires careful investigation. Such cases may create undesirable pathways. Therefore, it is necessary to assess the integrity of the cement and any possible leakage into the surrounding formations.

By carefully examining the cementing plan and related reports, the relevant engineer can identify the main sources of damage to the formation and take appropriate corrective measures. As a result, the comprehensive review of the cementing program, along with the evaluation of the relevant factors, plays an important role in reducing possible damage to the formation. Through careful monitoring and investigation, the relevant engineer can minimize the adverse effects related to the cementing operation and ensure reservoir integrity and efficiency.

1.4.7 Well Completion and Perforation Reports

A thorough review of well completion and perforation reports, along with an analysis of the fluids used in these operations, is critical to assessing the potential for formation damage. Well-completion operations and used fluids have the potential for damage. For example, perforating with contaminated fluids can cause severe damage. During this operation, solids are injected into the formation at high pressure and speed, causing severe damage to the production interval. Perforating in oil-based drilling mud can also cause an unfavorable change in wettability. The change in wettability and the transition from a water-wet to an oil-wet state can reduce the permeability of the production interval to oil and gas. The issue of formation damage caused by oil-based muds is considered to be very minor in most cases. Although it is believed that oil-based fluids do not cause damage, In addition, the presence of debris and chips from perforation operations is a major concern. Completion fluids are not always compatible with the minerals of the producing formation, and this issue is more important in water-sensitive sandstones than in carbonate reservoirs. Core testing of formation water should be carried out to ensure fluid compatibility. These tests can be considered when correcting drilling fluids and adding appropriate additives for water-sensitive reservoirs. In summary, a comprehensive review of well completion and perforation reports is required, with particular emphasis on fluid compatibility and the potential for formation damage. By conducting detailed assessments and implementing appropriate measures, such as core testing and proper acid design, professionals can reduce the risks associated with well completion and perforation operations, thereby ensuring optimal production layer performance.

1.4.8 History of the Well Workover

Workovers can cause formation damage in several ways. Workover reports should be analyzed to investigate potential damage, such as:

Contaminated Fluids: The use of impure fluids during workover operations can pose a risk of formation damage. These fluids may contain impurities or substances that can adversely affect the reservoir matrix.

Incompatibility of Well Workover Fluids with Produced Salt Water: The compatibility of the well workover fluid used during operations with the produced salt water from the well should be verified. Incompatibility can result in the formation of insoluble deposits that can adversely affect the reservoir.

Paraffin Precipitation: Paraffin precipitation can occur during workover operations due to the temperature drop. The paraffin deposits can impede fluid flow in the reservoir and adversely affect productivity.

Water Plugging: Workover operations can inadvertently cause water plugging, where water is trapped in the pore spaces of the formation. This can impede hydrocarbon flow and reduce reservoir performance.

Therefore, a thorough review of workover reports allows for the identification and evaluation of any damage that may have occurred during the operation.

1.4.9 Well Acidizing and Stimulation History

The stimulation history is an important source of valuable and useful information when analyzing well. The stimulation history should be examined to identify the causes of damage in each case. It can also be used to guide the design of future stimulations. Well-stimulation is the work performed on oil or gas wells to increase production by improving the flow of hydrocarbons from the reservoir into the well. This involves creating new pathways in the rock to facilitate oil and gas production. Well-stimulation is typically used when a well has experienced a decline in production. The effectiveness of well stimulation can be evaluated by performing well production tests before and after the operation. Analyzing changes in skin factor and monitoring the amount of acid injected during the operation can also provide useful information. In summary, a comprehensive evaluation of well-stimulation operations requires the integration of several evaluation techniques and a thorough analysis of the acidizing and stimulation records. This comprehensive approach increases the decision-making power for a successful acidizing operation and helps to optimize stimulation strategies by considering the economic factors of the well.

1.5 Types of Formation Stimulation Methods

The aim of well-stimulation methods is to have the appropriate effect on the damaged area. By restoring or increasing the permeability of this area to, or beyond, the optimum, the productivity index or injectivity of the well can be improved. Formation stimulation methods are discussed in detail in the following sections.

1.5.1 Acidizing

Acid injection aims to improve the connection between the formation and the well by dissolving parts of the reservoir rock, thus improving the well productivity and increasing the production rate at a constant reservoir pressure drop, or reducing the pressure or, more precisely, the energy loss at a constant production rate. Acid well stimulation methods are classified into the following categories based on the injection pressure and the penetration and dissolution power of the acid:

1.5.1.1 Acid Wash

This method is used to wash the screen channels and clean the wellbore entrance. This method involves injecting a small amount of acid followed by a neutral fluid such as diesel (to force the acid into the formation). The aim of this method is not to increase the permeability of the damaged area but to remove the skin factor caused by the perforation (Sp) and to clean the wellbore (Economides & Nolte, 1989).

1.5.1.2 Matrix Acidizing

In this process, acid is injected into the formation at a pressure lower than the fracture pressure of the formation. This helps fluid flow into the wellbore by dissolving the soluble impediments and making the pores bigger. In other words, the objective of this operation is to create oil or gas flow paths by dissolving the rock matrix, enlarging the existing fractures, and interconnecting the pores, thereby increasing the permeability and porosity percentage of the producing formation. This process can be used to stimulate both carbonate and sandstone formations. During matrix acidizing, the contact area between the acid and the formation is very large, so the frictional pressure increases as the volume and velocity of the acid increase. The most important phenomenon in matrix acidizing is the formation of a wormhole hole, which occurs in the vicinity of the borehole. Larger wormhole-shaped samples tend to receive more acid, which both expands their area of influence and travels a greater distance. This continuous impact creates a large channel in the rock. The multiple branches of the wormhole phenomenon depend on the injection velocity so that as it increases, a denser and thinner network of channels is created (Economides & Nolte, 1989; Kalatehno et al., 2023).

Example 1.4 What is acid wash, and when is it used?

Answer:

Acid wash is used to clean the wellbore entrance and wash the screen channels by injecting a small amount of acid followed by a neutral fluid, aiming to remove the skin factor caused by perforation.

1.5.2 Fracturing

Fracturing is one of the most effective and complex engineering techniques that can be used to improve the production capability of a well. Fracturing involves the injection of acid into the formation at a pressure higher than the fracture pressure of the formation. This type of stimulation creates a new fracture or increases the size of existing fractures, allowing oil or gas to flow out of the formation and into the well. This method can be used in carbonate reservoirs as a shortcut in the damaged area and also to stimulate undamaged and carbonated formations, which can have a favorable result. In this operation, the amount of acid penetration depends on the velocity of the acid, its rate of reaction with the formation, and the amount of contact of the fractures with the acid. The greatest penetration is achieved when a larger volume of acid is consumed the first time a fracture is created. Subsequent fractures caused by subsequent injections may have less penetration, cause the previous fractures to close, or disappear on their own when the pressure is released. In a naturally fractured formation, the acid velocity is determined by the injection rate. The deepest penetration occurs when the injection is made at a pressure close to that required to create an additional fracture in the formation, i.e., it should always be ensured that the optimum pressure is applied to create the first effective fracture before creating an additional fracture; therefore, if the pressure is higher than the optimum value, another fracture will be created, and as a result, the rate of acidizing injection will decrease. The degree of acid reaction has the greatest effect on the fracturing formation. In general, two types of fracturing formation operations are of interest and are discussed below.

1.5.2.1 Propped Fracturing (Hydraulic Fracturing)

Propped fracturing is currently the most widely used oil and gas well-stimulation technique in the world. In summary, once the necessary studies of the desired formation have been carried out, a low-viscosity fluid is first pumped into the well under pressure to create and widen a fracture. Once this fluid has been injected, the main fluid, which has a much higher viscosity than the previous fluid, is injected into the well along with the proppant. The main fluid increases the width of the fracture and extends it. The proppant also prevents the fracture from closing due to pressure from the top and side layers. Once all the fluids have been injected and pumping has stopped, the diluent in the main fluid reduces the viscosity and is removed. This leaves a high hydraulic conductivity channel for the oil to flow through (Yew & Weng, 2014).

1.5.2.2 Acid Fracturing

Acid fracturing is a well stimulation technique used in the oil and gas industry to enhance the productivity of reservoirs. It involves the injection of acid into the formation surrounding a wellbore to create or enhance fractures, allowing for better fluid flow and increased production rates (Gharibi & Zoveidavianpoor, 2015).

The process of acid fracturing begins with the selection of an appropriate acid solution based on the characteristics of the reservoir rock and fluids. Typically, hydrochloric acid (HCl) or mixtures of HCl with other acids such as formic acid are used. The acid solution is then prepared by diluting it with water to achieve the desired concentration (Al-Ameri & Gamadi, 2020).

Before injecting the acid, a thorough evaluation of the wellbore and reservoir is conducted to identify potential zones for treatment. This evaluation includes analyzing well logs, core samples, and production data to determine the optimal location and design for creating fractures. Once the treatment zone is identified, a perforating gun is used to create holes in the casing and cement sheath surrounding the wellbore. These perforations allow for direct access to the reservoir rock. Acid diversion techniques may also be employed to ensure that all sections of the formation are treated uniformly. The acid solution is then pumped into the wellbore under high pressure using specialized equipment such as pumps and surface control systems. The pressure helps propagate fractures in the formation by dissolving minerals present in the rock matrix or reacting with cementing materials. As the acid contacts the reservoir rock, it reacts with minerals such as calcite or dolomite, dissolving them and creating channels or pathways for fluid flow. This dissolution process enlarges existing fractures or creates new ones, increasing permeability and improving connectivity between reservoir zones (Hendrickson et al., 1992; McLeod, 1984).

To prevent excessive damage to nearby formations or unwanted fluid migration, various additives may be included in the acid solution. These additives can help control corrosion, reduce precipitation of solids, minimize clay swelling, or mitigate any potential environmental impacts. After a sufficient contact time between the acid and formation has been achieved, usually ranging from several hours to several days depending on factors like rock type and treatment objectives, flowback operations commence. Flowback involves recovering spent acid along with any produced fluids from within the fractured zone back to surface facilities for separation and disposal (Krueger, 1986).

Acid fracturing can significantly improve production rates by enhancing fluid flow pathways within a reservoir. It is commonly used in carbonate formations where matrix permeability is low but natural fractures exist. Acid fracturing can also be combined with other stimulation techniques like hydraulic fracturing (fracking) to maximize production potential. However, it's important to note that there are environmental considerations associated with acid fracturing due to potential groundwater contamination risks from chemical additives used in acids. Therefore, strict regulations are in place to ensure proper handling and disposal practices are followed during this process.

There are several correlations that can be used to estimate the productivity and performance of acid fracturing operations. Some commonly used correlations include:

1. Fracture conductivity correlation: This correlation relates the fracture conductivity to various parameters such as proppant concentration, proppant size, fracture width, and fluid viscosity. It helps in estimating the flow capacity of the created fracture.
2. Acid reaction rate correlation: This correlation estimates the rate at which acid reacts with the formation rock. It takes into account factors such as acid concentration, temperature, and rock mineralogy to predict the dissolution rate of minerals and subsequent increase in permeability.
3. Fracture geometry correlation: This correlation helps in estimating the dimensions of the created fracture, including length, width, and height. It considers parameters such as injection rate, fluid viscosity, rock properties, and treatment volume.
4. Proppant settling velocity correlation: Acid fracturing often involves injecting proppants into the created fractures to keep them open after acid treatment. This correlation estimates the settling velocity of proppants based on their size and density, which is important for designing an effective proppant placement strategy.
5. Acid penetration depth correlation: Acid penetration depth refers to how far into the formation rock the acid can effectively penetrate during fracturing operations. This correlation considers factors such as acid concentration, injection rate, rock permeability, and formation pressure to estimate the depth of acid penetration. It is important to note that these correlations are empirical in nature and may vary depending on specific reservoir conditions and acid fracturing techniques employed.

Example 1.5 What is the purpose of proppant in propped fracturing?

Answer:

In propped fracturing, the proppant is used to prevent the fracture from closing by maintaining pressure from the top and side layers, ensuring a high hydraulic conductivity channel for oil flow.

1.5.3 Thermal Stimulation

The use of steam to produce high-viscosity crude oil is now a widely recognized and successful technique. This technology is a new form of steam transfer in reservoirs where the fluid is essentially immobile at reservoir temperature due to its high viscosity. In this method, horizontal fractures are created in the production wells below the 5-spot pattern and then stimulated with high-pressure steam to heat the reservoir. A horizontal fracture is then created from the injection well to the production wells, and steam is injected at a very high rate and pressure in the first stage to keep the fracture open, heat the formation, and mobilize the viscous bitumen. In the second phase, the injection rate and pressure are reduced until the steam injection has a good effect and the liquefied bitumen moves towards the production wells. During this period, most of the bitumen production takes place. In the final stage, the ruby steam phase, the produced water is mixed with fresh water and circulated back into the reservoir, and bitumen production continues to the desired level. This process is a low-cost, energy-efficient method that also increases the production of extra-heavy oils (Fuhrberg, 1983).

Example 1.6 What is the process and benefit of thermal stimulation using steam?

Answer:

Thermal stimulation involves creating horizontal fractures and injecting high-pressure steam to heat the reservoir and mobilize viscous crude oil for production. It is energy-efficient, cost-effective, and suitable for producing extra-heavy oils.

1.6 Objectives of Matrix Acidizing

Matrix acidizing is a well-stimulation technique that involves injecting acid into the formation at pressures below the fracture pressure to dissolve and remove the damage and enhance the flow of hydrocarbons. Matrix acidizing is commonly used in carbonate and sandstone reservoirs, where the acid reacts with the minerals and creates wormholes or channels that increase the permeability and connectivity of the pore network.

The main objectives of matrix acidizing are:

- Improving well productivity: by increasing the effective wellbore radius and reducing the skin factor
- Enhancing reservoir performance: by improving the sweep efficiency and recovery factor
- Restoring formation permeability: by removing or bypassing the near-wellbore damage
- Economic considerations: by optimizing the cost–benefit ratio and minimizing the environmental impact

1.6.1 Improving Well Productivity

One of the primary objectives of matrix acidizing is to improve well productivity, which is defined as the ratio of the flow rate to the pressure drawdown at a given reservoir pressure. Well productivity depends on several factors, such as:

- **The reservoir permeability**: which determines the ease of fluid flow in the formation
- **The wellbore radius**: which determines the cross-sectional area for fluid flow in the wellbore
- **The skin factor**: which represents the additional pressure drop caused by damage or stimulation near the wellbore

Matrix acidizing can improve well productivity by increasing the effective wellbore radius and reducing the skin factor. The effective wellbore radius is the equivalent radius of a hypothetical wellbore that would produce at the same rate and pressure as the actual wellbore with damage or stimulation. The skin factor is a dimensionless parameter that quantifies the deviation of actual flow from ideal radial flow due to damage or stimulation. In carbonate reservoirs, matrix acidizing can create wormholes that extend deep into the formation and bypass the damaged zone near the wellbore. Wormholes are highly conductive channels that result from the preferential dissolution of carbonate minerals by acid. Wormholes can increase the effective wellbore radius significantly and reduce or even eliminate the skin factor.

In sandstone reservoirs, matrix acidizing can repair or dissolve damage from drilling, completion, workover, or production operations. Damage can include drilling solids, filtrate invasion, scale deposition, paraffin accumulation, asphaltene precipitation, emulsion formation, fines migration, clay swelling, and bacterial growth. Matrix acidizing can restore or enhance the formation permeability near the wellbore and reduce the skin factor (Chacon & Pournik, 2022; Kasi et al., 2022).

The improvement in well productivity due to matrix acidizing can be estimated by using various models and methods, such as:

- **The Radial Composite Model**: assumes that the formation consists of two concentric regions with different permeabilities representing the damaged and undamaged zones (Daolun et al., 2020).
- **The Pseudo-skin Model**: assumes that matrix acidizing creates a high-permeability zone around the wellbore that can be represented by a negative skin factor (Farokhi & Gerami, 2012).
- **The Wormhole Model**: assumes that matrix acidizing creates wormholes that can be represented by an equivalent radius or conductivity (Istchenko & Gates, 2014; Rahaman et al., 2016).
- **The Nodal Analysis Method**: which uses a systems approach to analyze and optimize the flow performance of a well-reservoir system

1.6.2 Enhancing Reservoir Performance

Another objective of matrix acidizing is to enhance reservoir performance, which is defined as the ability of a reservoir to produce hydrocarbons over time under various operating conditions. Reservoir performance depends on several factors, such as:

- **The Reservoir Pressure**: which determines the driving force for fluid flow in the reservoir
- **The Reservoir Heterogeneity**: which determines the distribution and connectivity of different rock types and properties in the reservoir
- **The Reservoir Sweep Efficiency**: which determines how effectively a displacement fluid contacts and displaces oil or gas in place in the reservoir
- **The Reservoir Recovery Factor**: which determines how much oil or gas can be recovered from a reservoir under a given recovery mechanism

In carbonate reservoirs, matrix acidizing can create wormholes that increase the heterogeneity and connectivity of the pore network. Wormholes can also improve the recovery factor by increasing the contact area between the injected fluid and the oil or gas in place. In sandstone reservoirs, matrix acidizing can improve sweep efficiency by restoring the natural flow paths and reducing the bypassed oil or gas (Chacon & Pournik, 2022; Kasi et al., 2022).

Utilizing a variety of models and techniques, one can estimate the improvement in reservoir performance due to matrix acidizing:

- **The Reservoir Simulation Model**: which uses numerical methods to solve the fluid flow equations in a discretized representation of the reservoir geometry and properties
- **The Material Balance Method**: which uses a mass conservation principle to relate reservoir pressure and production history to the original and remaining volumes in place in a reservoir
- **The Decline Curve Analysis Method**: which uses empirical or analytical functions to fit and extrapolate the production history and forecast the future performance of a well or reservoir
- **The Reservoir Engineering Analysis Method**: which uses analytical or semi-analytical solutions to model the fluid flow behavior and performance of a well or reservoir under various operating conditions

Example 1.7 What does improving the sweep efficiency in a reservoir signify?

Answer:

Improving the sweep efficiency means enhancing how effectively a displacement fluid contacts and displaces oil or gas in place, which leads to an increased recovery factor.

1.6.3 Restoring Formation Permeability

A third objective of matrix acidizing is to restore formation permeability, which is defined as the ability of a porous medium to transmit fluids under a given pressure gradient. Formation permeability is one of the key reservoir properties that affect the productivity and performance of a well or reservoir.

There are several factors that can reduce or impair formation permeability, including:

- **The nNatural Heterogeneity of the Formation**: which results from the variation and distribution of different rock types and properties in the formation
- **The Diagenetic Processes of the Formation**: these result from the physical and chemical changes of the rock matrix due to temperature, pressure, and fluid interactions over geological time
- **The Near-Wellbore Damage of the Formation**: which results from the invasion and deposition of foreign materials or fluids during drilling, completion, workover, or production operations.

Removal of formation damage or creation of high-permeability pathways can be achieved by matrix acidizing with appropriate acid systems and injection rates. The acid systems should be compatible with the formation fluids and matrix and have sufficient dissolving power to remove damage without causing adverse reactions or creating new damage. The injection rate should be low enough to prevent formation fracturing but high enough to create wormholes in carbonate reservoirs.

There are several ways to measure the matrix acidizing-related restoration of formation permeability, including:

- **The Core Analysis Method**: uses laboratory tests on rock samples obtained from the wellbore or formation to measure the porosity, permeability, mineralogy, and morphology of the rock matrix before and after acidizing (McPhee et al., 2015).
- **The Well Testing Method**: this uses field tests on reservoir samples obtained from the wellbore or formation to measure the pressure and flow rate of fluids in the wellbore and the formation before and after acidizing.
- **The Logging method**: uses various tools and sensors that are lowered into the wellbore or deployed on the surface to measure the physical properties of the formation and fluids along the wellbore before and after acidizing

1.6.4 Economic Considerations

A final objective of matrix acidizing is to consider the economic aspects of the stimulation technique. Economic considerations include the cost–benefit analysis, the risk-reward analysis, and the environmental impact assessment of matrix acidizing. The cost–benefit analysis is the evaluation of the costs and benefits associated with matrix acidizing. The costs include the expenses for the acid system, the additives, the equipment, the personnel, and the operation time. The benefits include revenues from the increased production rate, the improved recovery factor, and the extended well life.

1.7 Carbonate Reservoirs

A reservoir rock is a rock that can store and transfer fluids. The conditions and depositional environment determine the types of minerals in the reservoir rock, but generally, reservoirs can be divided into three categories: carbonate, sandstone, and mixed reservoirs. Mixed reservoirs consist of carbonate, sand, and shale layers. The shale layers weaken the connectivity between the layers and cause heterogeneity in fluid distribution due to the presence of different rocks in this type of reservoir.

Carbonate reservoirs are reservoirs that contain hydrocarbons in carbonate rocks, which are rocks composed of calcium carbonate minerals such as calcite and dolomite. Carbonate reservoirs are estimated to hold more than 60% of the world's oil and 40% of the world's gas reserves. Carbonate reservoirs are very common in the Middle East, where they account for 70% of oil and 90% of gas reserves (Kargarpour, 2020). Carbonate rocks often have more complex properties than sandstones. This type of reservoir consists of a network of interconnected low-porosity, high-permeability fractures adjacent to high-porosity, low-permeability matrices. The fracture network acts as a highway, transporting the produced fluid through the matrix to the wellbore at a faster rate. Initially, production occurs through the fractures, but due to the low pore volume of the fractures, little hydrocarbon fluid is produced. After some time, the matrix enters the production cycle, and hydrocarbon fluid flows from the matrix, which contains more oil, to the fracture and is produced (Kalatehno et al., 2023). Carbonate constituents typically consist of more than 90% limestone (calcium carbonate) and/or dolomite (magnesium calcium carbonate). One of the most important methods of increasing or improving well production is acid stimulation. Carbonate reservoirs inevitably have heterogeneities that make the predictions of models based on homogeneous formations very different from reality. Well, operations can damage the formation from the first moment of drilling until the well is abandoned. Diagnosing the cause of formation damage and developing methods to avoid or prevent it have been the subject of many studies. Blocking of the fractures or the rock structure significantly changes the performance of the well, which can be seen as a decrease in hydrocarbon production (Chabrelie et al., 2007).

To understand and develop carbonate reservoirs, it is important to use an integrated approach that combines different methods and data sources at different scales. Some of the methods that can be used to study carbonate reservoirs are well testing, well performance analysis, core analysis, fluid sampling and testing, and logging. These methods can provide information on the lithology, porosity, permeability, saturation, pore geometry, fracture density, fracture orientation, fracture aperture, fracture connectivity, fluid type, fluid pressure, fluid temperature, fluid viscosity, fluid density, and fluid compatibility of carbonate reservoirs. Based on this information, an optimal development plan can be designed and implemented to enhance the recovery of hydrocarbons from carbonate reservoirs (Kargarpour, 2020).

1.7.1 The Composition and Characteristics of Carbonate Reservoirs

The composition and characteristics of carbonate reservoirs depend on the depositional environment, the diagenetic history, and the tectonic setting of the carbonate rocks. Different reservoir properties can result from various carbonate reservoir compounds, but some common types of carbonate reservoirs are as follows (Kargarpour, 2020):

- **Organic Build-ups Including Reefs**: These are structures formed by the accumulation of organic material such as corals, algae, sponges, and bryozoans. They can create large mounds or platforms with high porosity and permeability in the core and low porosity and permeability in the flank.
- **Grainstone Shoals on Shelves**: These are sand-sized carbonate particles that form elongated ridges or bars on shallow shelves. They can have high porosity and permeability due to the sorting and winnowing of grains by waves and currents.
- **Subtidal and Intertidal Complexes:** These are carbonate sediments that form in the zone between low tide and high tide. They can have various textures and structures, such as mudstones, wackestones, packstones, oolites, stromatolites, and tidal flats. They can have variable porosity and permeability depending on the degree of compaction, cementation, and dissolution.
- **Karstification and Paleocave Systems**: Karstification and paleocave systems are features that result from the dissolution of carbonate rocks by acidic groundwater or surface water. They can create large cavities or networks of fractures that enhance the porosity and permeability of the reservoir. However, they can also be filled with sediments or minerals that reduce porosity and permeability.
- **Chalk**: This is a fine-grained carbonate rock composed of microscopic calcareous organisms such as foraminifera and coccoliths. It can have low porosity and permeability due to compaction and cementation, but it can also be fractured or dissolved to create secondary porosity and permeability.

Example 1.8 In what ways can matrix acidizing benefit sandstone reservoirs specifically?

Answer:

In sandstone reservoirs, matrix acidizing can repair or dissolve damage from other operations, restoring or enhancing the formation permeability near the wellbore and consequently reducing the skin factor.

Example 1.9 What type of carbonate reservoir contains structures formed by organic material build-ups?

Answer:

Organic build-ups including reefs are structures in carbonate reservoirs formed by the accumulation of organic materials like corals and algae.

1.8 Acid-Rock Interactions

Acid-rock interactions are the chemical reactions that occur between the injected acid and the formation rock during matrix acidizing. These reactions can alter the porosity, permeability, mineralogy, and morphology of the rock matrix and affect the transport and distribution of fluids in the pore space. The main factors that influence acid-rock interactions are:

- **The Type of Rock**: The type of rock determines the mineralogy and morphology of the formation and the solubility and reactivity of the minerals with acid.
- **The Type of Acid**: The type of acid determines the dissolving power and compatibility of the acid with the formation fluids and matrix.
- **The Concentration of Acid**: The concentration of acid determines the amount and rate of dissolution or precipitation of minerals by acid.
- **The Temperature of Acid**: The temperature of acid determines the reaction kinetics and thermodynamics of acid-rock interactions.
- **The Injection Rate of Acid**: The injection rate of acid determines the pressure and flow regime of acid in the formation and the extent and distribution of stimulation.
- **The Contact Time of Acid**: The contact time of acid determines the duration and depth of penetration or diversion of acid in the formation.
- **The Presence of Additives**: The presence of additives determines the corrosion rate, reaction rate, chelation ability, adsorption behavior, dispersion effect, emulsification tendency, and environmental impact of acid.

The main outcomes of acid-rock interactions are changes in the porosity, permeability, mineralogy, and morphology of the rock matrix and the pore space. These changes can be positive or negative, depending on the type and extent of dissolution or precipitation of minerals by acid. For example, porosity and permeability can increase due to the dissolution of minerals or the removal of fines or decrease due to the precipitation of minerals or the plugging of fines. Similarly, mineralogy and morphology can change due to the dissolution or precipitation of minerals or the alteration of rock texture or structure (Dankwa et al., 2016; You & Lee, 2021).

1.9 Types of Acids in Acidizing

Acidizing is a technique used to increase the production rate of oil wells by injecting a selected acid and removing the damage in the well. The choice of acid depends on various factors, such as the type of reservoir damage, rock type, temperature, reservoir pressure, and economic considerations. This part reviews the types of acids commonly used in acidizing (Wang, 2021).

1.9.1 Mineral Acids

- **Hydrochloric Acid (HCl)**

Hydrochloric acid (HCl) is a mineral acid that is most commonly used in carbonate rocks. It has a high capacity to dissolve minerals such as calcium and magnesium that are common in these rocks. The typical concentration of HCl used in acid treatment is 15% by weight (Bazin, 2001; Hendrickson et al., 1992). However, HCl also has strong corrosive properties that can react with metal surfaces to form ferrous deposits. To prevent this, a corrosion inhibitor must be used (Buijse et al., 2003; Hendrickson et al., 1992; Li et al., 2008; McLeod, 1984; Mumallah, 1991; Schechter, 1992). The reactions of calcite and dolomite with HCl are shown below:

$$2HCl + CaCO_3 \rightleftarrows CaCl_2 + H_2O + CO_2$$

$$4HCl + CaMg(CO_3)_2 \rightleftarrows CaCl_2 + MgCl_2 + 2H_2O + 2CO_2$$

HCl can be used in different concentrations depending on the purpose of the acid treatment. Lower concentrations (5–15% w/w) are used to remove sediments, salt plugs, and emulsions, while higher concentrations (15–37% w/w) are used to create longer and wider paths in the reservoir rock (Fredd & Fogler, 1998, Al-Anazi et al., 2003, Al-Muhareb et al., 2003, Nasr-El-Din et al., 2003, NASR-EL-DIN et al., 2004, Fan et al., 2005).

1.9.2 Organic Acid

Organic acids are a type of acid that is less corrosive and more stable at high temperatures than mineral acids. They are suitable for situations where the acid needs to be in contact with equipment and pipes for longer periods or where the acid has to travel longer distances to reach the reservoir. Organic acids can dissolve more minerals than mineral acids, but they are also more expensive and should only be used when necessary (Al-Saif et al., 1975, Buijse et al., 2003, Al-Harbi et al., 2013,

Noshi, 2015, Alhamad et al., 2020). Some of the common organic acids used in acid treatment are:

- **Acetic Acid**

Acetic acid is an organic acid with the molecular formula CH_3COOH. It is more expensive and less corrosive than HCl and formic acid. The reactions of calcite and dolomite with CH3COOH are shown below (Nasr-El-Din et al., 2001; Wang et al., 2021):

$$2CH_3COOH + CaCO_3 \rightleftarrows Ca(CH_3COO)_2 + H_2O + CO_2$$

$$4CH_3COOH + MgCa(CO_3) \leftrightarrows Mg(CH_3COO)_2 + Ca(CHCOO)_2 + 2H_2O + 2CO_2$$

An advantage of using acetic acid is that it prevents the precipitation of ferric iron and the formation of $Fe(OH)_3$ by keeping ferric iron in solution.

- **Formic Acid**

Formic acid (HCOOH) is an organic acid widely used in the acidizing of oil and gas wells. This process aims to increase well productivity by injecting acid solutions into the well to dissolve formation damage. The reaction of formic acid with carbonate rocks such as calcite and dolomite produces calcium formate, carbon dioxide, and water, as shown below:

$$2HCOOH + CaCO_3 \leftrightarrows Ca(COOH)_2 + H_2O + CO_2$$

Formic acid has several advantages over mineral acids such as hydrochloric acid (HCl). It can prevent the precipitation of ferric iron compounds, which can cause fouling or corrosion. It can also reduce the corrosion rate of HCl at high temperatures. Formic acid can be used as a single acid for carbonate formation or as a base acid in combination with other acids such as hydrofluoric acid (HF) or acetic acid. The concentration of formic acid depends on the purpose of the acid treatment. Lower concentrations (5–15 wt%) are used to remove scale, salt plugs, and emulsions, while higher concentrations (15–37 wt%) are used to create longer and wider paths through the damaged areas (Daumer et al., 2010; Wang et al., 2021).

- **Citric Acid**

Citric acid ($C_6H_8O_7$) is a weak organic acid that plays an important role in the oil and gas industry. It is used in acidizing techniques, which are methods of increasing the productivity of hydrocarbon reservoirs by injecting fluids that dissolve carbonate minerals in the formation. This creates pathways for the easier flow of hydrocarbons. Citric acid is suitable for use in carbonate formations where iron deposits can form as a result of acid contact with equipment and pipes. Citric acid can prevent the formation of these deposits in hydrochloric acid solutions.

1.9.3 Mixed Acid

Mixed acids are a type of acid consisting of a combination of inorganic and organic acids. The composition of mixed acids varies depending on the application of the acid treatment. The main advantage of using mixed acids is that they can reduce the rate of reaction of the acid with the rock, which can increase the depth of penetration of the acid into the formation. Mixed acids are also suitable for high-temperature formations where single acids may not be sufficiently stable (Dai & Ben Mahmud, 2019; Zhao et al., 2018).

- **Mud Acid (HF-HCl)**

Mud acid (HF-HCl) is a type of mixed acid used in sandstone rock. It contains hydrochloric acid (HCl) and hydrofluoric acid (HF), which can dissolve quartz minerals that are resistant to HCl alone. The typical concentration of mud acid is 12% HCl and 3% HF by weight for sandstone rocks (Smith & Hendrickson, 1965). However, hydrofluoric acid loses its effectiveness at temperatures above 200 °F because it is consumed very quickly (Al-Harthy et al., 2009). Mud acid has a similar corrosivity to HCl, and the same corrosion inhibitors can be used for both. HF can also produce many by-products as a result of its reaction with reservoir rock. These by-products are listed in the Table 1.1.

Example 1.10 What is the purpose of using mud acid, a combination of HCl and HF, in sandstone formations?

Answer:

Mud acid, which combines HCl and HF, is used to dissolve quartz and other minerals resistant to HCl alone, thereby increasing permeability.

- **Emulsified Acid**

Emulsified acid is a type of mixed acid consisting of an aqueous acidic phase, usually hydrochloric acid (HCl), and an oil phase, usually diesel or kerosene, emulsified with

Table 1.1 HF reaction by-products

Name	Formulation
Ortho silicic acid	H_4SiO_4
Calcium fluoride	CaF_2
Sodium fluorosilicate	Na_2SiF_6
Sodium fluoro aluminate	Na_3AlF_6
Potassium fluorosilicate	K_2SiF_6
Ammonium fluorosilicate	$(NH_4)_2SiF_6$
Calcium fluorosilicate	$CaSiF_6$
Aluminum fluoride	AlF_3
Aluminum hydroxide	$Al(OH)_3$
Iron sulfide	FeS

a surfactant. It is used in the acidizing in the oil and gas industry, particularly for carbonate reservoirs that have high reactivity with acid (Shafiq et al., 2019).

Emulsified acid can have different compositions and concentrations depending on the type of formation and the desired outcome of the acidizing (Hoefner & Fogler, 1987; Navarrete et al., 2000; Adewunmi et al., 2022). Some common types of emulsifying acid are:

- **Acid-in-Oil Emulsion**: This is an emulsion in which the acid (water) phase is dispersed in the oil phase. It has greater stability and viscosity than an oil-in-acid emulsion and can produce better penetration and wormholes. It can also be used for Fracturing and acidizing up to 350 °F.
- **Oil-in-Acid Emulsion**: This is an emulsion in which the oil phase is dispersed in the acid (water) phase. It has lower stability and viscosity than an acid-in-oil emulsion and may have better solubility and compatibility with other additives.
- **Inverse Emulsion**: This is an emulsion in which the continuous phase changes from oil to acid or vice versa depending on temperature, pressure, or pH conditions. It can provide better compatibility and flexibility for different reservoir conditions.

- **Gelled Acid**

Gelled acid is a type of mixed acid consisting of an aqueous acid phase, usually hydrochloric acid (HCl), and a polymer phase that increases the viscosity of the acid. It is used in the acidizing in the oil and gas industry, particularly for carbonate reservoirs that have high reactivity with acid (Wang et al., Liu et al., 2021; Sui et al., 2022). Gelled acid can have different compositions and concentrations depending on the type of polymer (Chang et al., 2001; Crowe et al., 1981; Gomaa & Nasr-El-Din, 2010a, 2010b; Lungwitz et al., 2007; Taylor & Nasr-El-Din, 2003). Some common types of gelled acid are:

- **Linear Polymer Gels**: These are gels containing a linear polymer, such as polyacrylamide, which is soluble in HCl. They have a medium viscosity and can cause some delay and deflection.
- **Cross-linked Polymer Gels**: These are gels containing a cross-linked polymer network formed by the reaction of a polymer such as polyacrylamide and a cross-linking agent such as iron, zirconium, or aluminum ions. They have high viscosity and can produce significant penetration depth and deflection.
- **In Situ Gelled Polymers**: These are gels formed in situ by the cross-linking reaction between a polymer such as polyacrylamide and a multivalent metal cation such as iron, zirconium, or aluminum ions. They have variable viscosity and can provide dissolution and deflection properties due to the sudden increase in viscosity when exposed to acid.
- **In Situ VES Gels**: These are gels that viscoelastic surfactant (VES) molecules form on the spot. They have variable viscosity and can provide

- **Retarded Acid**

Retarded acid is a type of mixed acid that is made up of an aqueous acidic phase, usually hydrochloric acid (HCl) or hydrofluoric acid (HF), and a retarding agent that slows down the rate at which the acid reacts with the reservoir rock. It is used in the acidizing in the oil and gas industry, particularly for carbonate and sandstone reservoirs that have high reactivity with acid (Fayzi et al., 2015; Lian et al., 2021; Mahmoud et al., 2020). Retarded acid can have different compositions and concentrations depending on the type of retarder used (Mahmoud et al., 2020). Some common retardants are:

- **Aluminum Chloride**: This is a salt that can form a complex with HF and retard its reaction with silicate minerals. It can also reduce the rate of corrosion of metals and equipment.
- **Boric Acid**: This is an organic acid that can buffer the pH of HF and slow its reaction with silicate minerals. It can also prevent the precipitation of iron and calcium ions.
- **Phosphonic Acid**: This is an organic acid that can buffer the pH of HF and retard its reaction with silicate minerals. It can also prevent the precipitation of iron and calcium ions.
- **Retarded Acid System**: This is a mixture of HCl and an organic acid such as acetic acid or formic acid, retarded with an additive such as a polymer or ester. It can retard the rate of reaction of HCl with carbonate minerals such as calcite and dolomite, allowing deeper acidizing of the formation.

1.10 Reservoir Permeability and Porosity

Reservoir permeability and porosity are two fundamental reservoir properties that relate to the amount of fluid contained in a reservoir and its ability to flow. These properties have a significant impact on petroleum field operations and reservoir management. Permeability is a measure of the ease with which fluids can flow through a porous medium under a given pressure gradient. Permeability is usually expressed in darcies or millidarcies (md). Permeability depends on the size, shape, and distribution of the pore channels in the rock, the type and number of fluids present, the fluid flow rate, the length and cross-sectional area of the rock, and the pressure differential across the length of flow. Permeability can be measured by laboratory tests on core samples or by well tests on reservoir samples. Porosity is a measure of the void space in a porous medium that can be occupied by fluids. Porosity is usually expressed as a fraction or percentage of the total rock volume. Porosity depends on the size, shape, and arrangement of the rock grains, and the degree of compaction, cementation, and dissolution of the rock matrix (Che et al., 2023; Dargi et al., 2023).

1.11 Factors Affecting Acidizing Success in Carbonate Reservoirs

Acidizing is a reservoir stimulation technique that involves injecting acid into the formation to dissolve and remove the damage and enhance the flow of hydrocarbons. Acidizing is commonly used in carbonate reservoirs, where the acid reacts with the carbonate minerals and creates wormholes or channels that increase the permeability and connectivity of the pore network. The success of acidizing in carbonate reservoirs depends on various factors, such as:

- **Acid Properties and Compatibility**: which determine the dissolving power and stability of the acid system
- **Formation Temperature and Pressure**: which determine the reaction kinetics and thermodynamics of acid-rock interactions.
- **Reservoir Fluid Composition**: which determines the fluid-rock and fluid–fluid interactions and compatibility
- **Wellbore Conditions**: which determine the injection rate and pressure of acid and the distribution and diversion of acid in the formation

Example 1.11 What factors influence the success of acidizing in carbonate reservoirs?

Answer:

The success depends on factors such as acid properties and compatibility, formation temperature and pressure, reservoir fluid composition, and wellbore conditions which determine acid distribution within the formation.

1.12 Parameters Affecting Acidizing Operation

The rate and depth of acid penetration are strongly influenced by the acid injection rate. A high injection rate can reduce the reaction rate and acid consumption, allowing the acid to travel further and affect a larger area (Wang et al., 1993). In matrix acidizing, the objective is to remove the damage and deposits that impede fluid flow. Therefore, the acid injection pressure should be lower than the fracture pressure so as not to create cracks and fractures in the formation that would reduce the efficiency of the acidizing. The fracture pressure can be determined from geomechanical tests, leak-off tests, and drilling data (Rabbani et al., 2018). In any acidizing operation, the acid should not create insoluble deposits that would impede fluid flow in the porous medium. Therefore, the reaction of the acid with the formation minerals and the hydrocarbon fluid should be studied before any acidizing. Knowledge of the mineralogical composition of the formation and the composition of the hydrocarbon fluid is essential in selecting the best acidizing fluid. Formation pressure and temperature have a significant effect on the rate of reaction and dissolution of minerals.

Accurate knowledge of these parameters will help in the selection of the type of acid and its concentration (Khuzin & Khizhniak, 2019).

To acidizing carbonate rocks in the laboratory, a core sample of the desired formation is required. This type of rock has unique characteristics, such as a fracture network, which make it impossible to prepare a synthetic core in the laboratory according to the rock mineralogy. Carbonate reservoirs are heterogeneous reservoirs with different porosity and permeability at different locations. This difference causes the distribution and saturation of fluids to vary at different locations. During acid injection, this heterogeneity can lead to uneven acid distribution across the formation. This uneven distribution reduces the efficiency of acidizing. Therefore, characterizing the formation and dividing it into different zones with similar properties improves the acidizing process (Jafarpour et al., 2021).

1.13 Matrix Acidizing of Carbonate Formations

Using the matrix acidizing technique, it is possible to repair drilling and production-related damage that reduces the reservoir rock's natural permeability. Drilling fluids, cement, and other contaminants that obstruct the rock's pores and fractures may be the cause of the damage. It can also be caused by mineral deposits and organic matter that accumulate over time. The location and severity of the damage will affect the success of matrix acidizing (Dargi et al., 2023; Harry & Mcleod, 1984). Matrix acidizing of carbonate rocks is a specific type of matrix acidizing that targets the damaged areas close to the wellbore by injecting acids at pressures below the fracture pressure of the formation. Matrix acidizing of carbonate rocks has been extensively studied for several decades using experimental and mathematical methods to understand its complex details. It is still used in high-permeability formations with significant damage (50 mD or more). However, in low-permeability carbonate formations (less than 10 md), fracturing is usually preferred as it can provide more stimulation by creating long fractures in the reservoir. Matrix acidizing of carbonate reservoirs is an effective technique for enhancing well performance by injecting an acidic solution into the formation. This technique, also known as matrix acidizing of carbonate reservoirs, uses strong acids to stimulate the formation (Bazin, 2001; Palharini Schwalbert et al., 2020; Wang, 2021). Sedimentary rocks are a diverse group of rocks ranging from pure calcite and dolomite to clay-free sandstone. Dolomitization can increase porosity but decrease permeability by precipitating dolomite crystals. Unlike sandstone acidizing, carbonate acidizing is not usually aimed at removing formation damage but at reducing it from the initial level. The main difference between carbonate and sandstone acidizing is the composition of the acid solution. If the solubility of the rock in HCl is less than 20%, the HCl-HF acidizing method is used for sandstone rock. Chalk and marl formations, which are mainly composed of calcite and dolomite, are soluble in HCl alone and are suitable candidates for carbonate acidizing where HCl is used without HF. Carbonate acidizing is less complicated than sandstone acidizing because it is less prone to sedimentation.

HCl is usually used for carbonate acidizing. At temperatures above 400 °F, organic acids can be used instead to prevent corrosion. The purpose of acidizing with HCl is to remove the damage by creating high-permeability channels or to dissolve the blocked fractures in low-permeability fractured formations. The products of calcite reacting with HCl, CO_2, and $CaCl_2$, are completely soluble in water. This means that deposits of $CaCl_2$ and CO_2 are not a problem. Even if $CaCl_2$ precipitates or CO_2 separates into a gas phase, these phases will be dissolved with the continued production of oil, gas, or water. Contrary to popular belief, modeling acid with HCl is a difficult task. The reason for this difficulty is the slow reaction rate compared to the reaction rate between HF and common minerals in sandstones. It is also important to note that the reaction rate of HCl with dolomite is slower than with calcite. As a result of the higher reaction rate, HCl tends to penetrate through the pathways in carbonate rocks, i.e. high-permeability fractures, and has less tendency to move with a uniform front through the formation (as in the acidizing of sandstone). As a result of the acid reacting with their walls, these pathways widen and become wormholes, much larger in diameter than normal pores. This process continues until a few large wormholes dominate so that the injected acid flows through these wormholes and widens them. The tendency to form large-diameter wormholes makes HCl acidizing difficult to model. Because the wormholes are large, they are considered the most important feature of acidizing. The success of carbonate acidizing depends on the formation of these preferential flow paths extending from the wellbore or perforations. When these wormholes pass through the damaged areas, the production fluid can be produced with a relatively low-pressure drop due to the large size of these channels compared to natural pores (Haias and Jaf). Therefore, the physical principles of carbonate acidizing can be divided into three parts:

- Determining the characteristics of the pores and wormholes created by the acid
- Determining the conditions under which these wormholes are formed.
- The rate of formation of these wormholes

The interaction and flow of acid and rock in the porous medium can occur in two ways:

1. The reaction between acid and rock is so fast that the acid moves along the pores.
2. The second mechanism proposed here is multi-branching. This occurs because the porous medium has interconnected pores of different sizes. Therefore, the acid can enter not only the larger pores but also the smaller ones. This leads to the formation of branches called wormholes in the rock. These wormholes depend on the injection rate of the acid and the reaction rate of the rock. The lower the injection rate, the more acid penetrates different branches and a larger wormhole is formed (Glasbergen et al., 2005).

Example 1.12 Why is modeling acidizing with HCl considered a difficult task?

Answer:

HCl tends to penetrate through existing high-permeability pathways, creating large-diameter wormholes, and does not move uniformly through the rock matrix, making it challenging to predict the acidizing outcomes accurately.

1.14 Field Applications

Keihani et al. (2024) conducted a ground breaking study on matrix acidizing, specifically addressing gas well acidizing in southern Iran. Their research focused on introducing an innovative three-stage approach utilizing hydrochloric acid (HCl) and VDA systems to enhance well productivity. What set their study apart was the integration of VDA throughout laboratory experimentation, simulation modeling, and operational execution, a departure from previous research methodologies. The authors meticulously detailed the development and optimization of acid solutions, notably 28% HCl and 15% VDA, tailored for the specific well under investigation. A notable aspect of their work was the examination of various additives, including corrosion inhibitors, non-emulsifiers, iron-sequestering agents, surfactants, and H_2S scavengers. The authors provided transparent insights into the selection criteria and concentrations of these additives. The results showcased a significant reduction in skin effect, from 5 to -1.89, post-acid injection. Field operation outcomes demonstrated a remarkable 100% increase in gas production, an 84% rise in gas pressure, and a reduction in BS&W from 7 to 3%. The comprehensive nature of their study, encompassing laboratory development, simulation modeling, and field operations, provided a holistic understanding of the VDA system's application in gas wells. The authors further distinguished their work by disclosing specific details about the VDA composition, enhancing transparency in the scientific discourse (Keihani et al., 2024).

Shajari and Malhani (2023) reported the challenges faced in traditional matrix acidizing operations in certain Iranian oil fields with low reservoir pressure, and the application of a new emulsified acid system, the Triphase Emulsified Acid (3PEA), in low-pressure carbonate reservoirs in the south of Iran. They stated that conventional matrix acidizing jobs had limited success in improving oil production rates, especially in depleted oil fields, due to the shallow acid penetration radius and the insufficient treated area around the wellbore. They introduced the use of nitrogen in stimulation treatments, creating a triphase emulsified acid system that included HCl, gasoil, and nitrogen, which aimed to increase the stimulated radius around the wellbore. They presented case studies from Field-M in southwest Iran, analyzing the performance of three acidizing scenarios in 15 oil wells, completed cased hole with perforated intervals of less than 50 ft. They provided detailed information on each well, such as drilling history, lithology, and injectivity before matrix acidizing. They designed the acidizing job with three stages: pre-flush, main acid injection with 3PEA, and post-flush, followed by a displacing fluid. They compared the results of conventional acid systems with low volume 3PEA and high volume 3PEA, and showed that the 3PEA system demonstrated superior performance, with a 78% improvement in skin factor and a 40% increase in production rate. They concluded that the triphase emulsified acid system was effective in carbonate reservoirs, especially in wells with shorter perforated intervals. They recommended acid volumes of 150 US Gal/ft for HCl 28% and 300 US Gal/ft for 3PEA. They reported that the application of 3PEA resulted in over 100% increase in production rate and flowing pressure, and an 80% reduction in skin factor (Shajari & Malhani, 2023).

Liu et al. (2023) reported their achievements in a challenging HP/HT exploratory reservoir in North Kuwait, where they tested the first well that showed commercial potential. They used an openhole multistage completion and acid fracturing treatment to target the Middle Marrat reservoir, which had high concentrations of H_2S and CO_2 in the reservoir fluid. They faced numerous drilling challenges, such as diverse formations and loss control, and drilled the well to 19,000 feet with a maximum inclination of 57°. They used a 15,000-psi openhole multistage completion system with closable fracturing ports made of corrosion-resistant alloy tubing. After the acid fracturing treatment, the well-produced more than 3000 B/D of oil, a significant improvement over offset vertical exploration wells in the same field. They outlined the challenges of drilling and well intervention in ultra-sour conditions with high H_2S levels, and the need for specialized equipment and materials. The authors also discussed the importance of the multistage completion design, which could withstand extreme fracture stimulation pressures and HPHT conditions. They detailed the completion design criteria, such as corrosion-resistant alloy equipment, metal-to-metal connections, and the use of H_2S scavenger fluid during testing. They demonstrated the success of the three-stage acid fracturing treatment and well testing in a high H_2S environment, which provided a production benchmark for developing the Middle Marrat reservoir in North Kuwait's exploratory fields. They concluded by emphasizing the significance of the findings for future field development strategies and the ongoing feasibility study for long-term production testing with specialized equipment (Liu et al., 2023).

Alvarez et al. (2022) reported their novel foamed-acid system for matrix stimulation of challenging low-pressure, highly-fractured limestone formations, especially those near the WOC. They implemented the system through CT to address uneven acid treatment caused by conventional foamed acids' short stability time. They used stable foaming acids with qualities ranging from 60 to 80%, which enabled successful treatment placement and eliminated the risk of acid segregation into water zones. They focused on carbonate reservoir stimulations in Northern Iraq, traditionally treated with conventional HCl. They applied foamed matrix stimulation treatments using the new acid system in wells with high risks of early water breakthrough. Despite the proximity to water zones, the foamed acid technology showed impressive results, leading to a production gain of up to 3000 BOPD without an increase in water production. They highlighted the benefits of the new foamed-acid system, such as reduced hydrostatic pressure, optimized placement in depleted formations, and self-diversion capabilities. They emphasized the applicability of the technology in various well conditions, such as naturally fractured carbonates, low reservoir pressure, challenging completions, and high contrast permeabilities. They incorporated detailed case studies (Case A, Case B, and Case C), which demonstrated the successful application of the foamed-acid technology in overcoming specific challenges, such as poor water quality, water influx zones, and low reservoir pressure. The results indicated improved well performance, increased productivity, and decreased flowback periods.

They asserted that the new foamed-acid system was a cost-effective and efficient solution for matrix stimulation in challenging carbonate formations. The technology not only enhanced oil production but also prevented water breakthrough, making it a valuable tool for reservoir management and stimulation programs (Alvarez et al., 2022).

Elsheikh et al. (2022) explored the efficiency of organic acids in stimulating the Sarah sandstone formation in Saudi Arabia, aiming to find alternatives to HCl acid, commonly used for matrix acidizing in sandstone formations under HP-HT conditions. They tested the injection of organic acids, specifically $C_2H_4O_2$-HF and $C_2H_2O_4$-HF, on Berea cores to determine the optimal acid volume, guided by the mineralogy of the Sarah formation, identified through XRD and SEM. They found that these acids, when used as a preflush and main treatment, could recover the initial permeability of damaged sandstone cores. They examined three acid combinations: mud acid (HCl-HF), oxalic-HF, and acetic-HF. They revealed that oxalic-HF acid was particularly effective in recovering the initial permeability of the Sarah sandstone formation, offering a potential alternative to conventional mud acid. They also emphasized the ability of organic acids to be employed in FASSA and hydraulic fracturing operations. They concluded by recommending the use of oxalic-HF acid for acidizing the Sarah sandstone formation, showcasing the potential of organic acids as a viable and corrosion-mitigating solution in HP-HT reservoir conditions (Elsheikh et al., 2022).

Al-Rekabi et al. (2020) discussed carbonate matrix acidizing design for improving the wormhole network and preventing damage caused by treatment fluids in Majnoon oilfield. They optimized the treatment fluid recipes to avoid sources of damage, such as sludge, emulsion, and wettability alteration, and maintain the newly acquired permeability during the acid injection phase. They also introduced a real-time skin trend monitoring model that uses wellhead pressure and flow rate values to interpret the formation response during acid injection and flow back clean-up phases. They applied an injection multi-rate drawdown well test model and a forward multi-rate drawdown test model to monitor the skin trend during the injection and clean-up phases, respectively, and considered the permeability contrast to monitor the diversion effect. The authors reported that the skin trend at the end of the flow back phase was significantly higher than the skin value at the end of the injection phase, indicating induced and added skin value. They defined "Stimulation Efficiency" as the ratio of Productivity Index at the end of clean-up to the index at the end of injection phase, and "Acidizing Induced Skin" value as the difference in skin values at the end of acid injection and clean-up phases. They found that the average added skin value ranged from + 3 to + 10, with an average stimulation efficiency of 30–40%. The authors described the methodology of using transient radial flow with multi-rate from a line source (Ei-Function) to obtain the skin factor for both injection and flow back periods. They evaluated drawdown tests to quantify reservoir flow characteristics, such as skin factor and dynamic permeability. They also considered various factors, such as hydrostatic pressure, frictional losses, tubing roughness, and the use of a self-degradable, polymer-based mechanical diverter. The authors concluded that the added skin value, ranging from + 3 to + 10, led to an average stimulation efficiency of

30–40%, resulting in an increase in skin value compared to the acid injection phase. They suggested potential production gain losses due to prolonged soaking periods, non-optimum volumes for non-emulsifier, inadequate displacement of pickling fluid, and the use of existing completion brine for new wells. They proposed mitigation methods, such as reducing soaking periods, adding surface tension reducers in post-flush fluids, and using 5–10% acid for pickling fluid to enhance the effect of iron control (Al-Rekabi et al., 2020).

Gao et al. (2019) reported their novel matrix stimulation technique, "Large Liquid Volume Deep Penetration Acidizing," for wells in Iraq's Missan oilfield, which faced challenges of low production performance and decreased formation energy. They proposed a two-fold increase in injection volume compared to conventional methods, based on the Darcy Productivity Equation, and used acid systems with specific characteristics. They introduced two acid systems, PA-VES and PA-GL, which were suitable for high permeability formations and had economic benefits of gelled acid. They showed that PA-VES, a self-diverting acid, had superior performance in viscosity, corrosion inhibition, and leakoff reduction. They also used additives like PA-OS3, an organic plugging remover, to clean organic scales before acidizing. They discussed the critical aspects of temperature calculation, injection rate optimization, and injection volume optimization. They showed simulation results that highlighted the importance of higher pump rates in reducing formation temperature and acid rock reaction rates. They presented three case studies that demonstrated the successful application of the technique in wells A, B, and C, achieving significant production enhancements. They concluded by emphasizing the importance of optimizing injection rates and liquid volumes, reducing leakoff, and managing temperature in promoting effective wormhole extension (Gao et al., 2019).

Letichevskiy et al. (2017) presented a novel technology for enhancing oil recovery in depleted carbonate reservoirs in Samara, Russia. They explained the challenges faced by the oil fields in the region, such as declining production, increasing watercut, low reservoir pressure, formation heterogeneity, and water–oil contact proximity. They introduced foam acid treatment as a promising solution, which was implemented in 2016. This technology uses a stable, non-damaging agent to create a 65% nitrogen foam at the bottom hole, which improves well flow back and restores natural flow. They reported remarkable results, such as a 69% greater initial well productivity index, a doubled starting production rate, and a 25% reduction in acid injection volume, compared to other methods. They stated that nitrogen-foam acid treatments are now a regular stimulation program, which has proven effective in challenging geological conditions and has contributed to enhanced oil recovery in Samara's mature fields. The authors also described the unique features of the foam acid treatment technology, which utilizes nitrogen-foam diverting to overcome challenges such as reservoir heterogeneity, water–oil contact proximity, and low-quality cementing. They demonstrated its superiority over conventional matrix acidizing through laboratory testing, illustrated by the Kuleshovskoye field case. They showed that over a six-month period, wells treated with foam had 38% higher normalized productivity index and 58.6% more cumulative production than those treated with standard methods. They concluded that the application of this technology promises continued success in optimizing production and costs in mature oil fields (Letichevskiy et al., 2017).

Zhou et al. (2009) reported their ground breaking diverting acid stimulation treatment technique for carbonate reservoirs in China, which posed challenges due to their deep burial, serious heterogeneity, and complex formations. They proposed a technique based on a self-diverting acid system with a visco-elastic surfactant, which showed significant improvements over traditional acid stimulation methods using polymer-based systems. They used a surfactant that formed micelle structures with acid, which increased viscosity in carbonate formations and prevented fluid leakage into pores, cracks, and wormholes. They applied the technique in five carbonate oil fields in China, with 38 wells subjected to acid fracturing and 25 wells treated, resulting in excellent outcomes. They delved into the acid treatment/fracture mechanism of in-situ self-diverting acid, which contained a viscoelastic surfactant and formed baculate micelles during the reaction with rocks. This led to a rapid increase in viscosity, enabling uniform treatment in heterogeneous formations. They conducted laboratory experiments, simulating the reaction between in-situ self-diverting acid and carbonate rock, and revealed a significant increase in viscosity during the reaction. They injected self-diverting acid into cores with different permeabilities and observed higher injecting pressure, indicating effective diversion. They compared the self-diverting acid with traditional acid and showed that the self-diverting acid outperformed traditional acid in terms of injecting pressure and volume, highlighting its diverting efficiency. They showed that the cores from different permeability groups had excellent diversion effects, with wormhole lengths accounting for a significant portion of the total length. They presented field cases, such as Well A in Qinghai oil field, which demonstrated improved production levels after stimulation. They showed that the self-diverting acid system demonstrated superior viscosity, diverting pressure, and penetration distance compared to traditional acid methods (Zhou et al., 2009).

Garcia et al. (2008) reported a groundbreaking acid-diversion technique employed in the Cantarell Field, the second-largest oilfield globally, located offshore in the Gulf of Mexico. They used a novel acid diverter based on associative polymer technology (APT) for approximately 55 high-permeability wells in highly fractured carbonate formations, which faced challenges due to heterogeneous wells and varying water saturations. They aimed to reduce water permeability without affecting hydrocarbon permeability significantly, and to optimize hydrocarbon recovery, which had evolved from 1.1 million B/D in 1981 to 2.1 million B/D in 1996. They compared APT with conventional methods, such as foam and in-situ crosslinked acid, which showed limitations and resulted in uneven acid distribution. They demonstrated APT's consistent and effective acid diversion, which was not restricted to high water saturation zones, through laboratory tests and production logs. They also showed the pressure response during APT acid jobs, which indicated efficient diversion. They concluded with a comparison of productivity index data, which highlighted the superiority of APT in enhancing well productivity over foam and in-situ crosslinked acid.

They suggested that APT offered a more reliable and consistent acid-diversion technique, addressing challenges posed by heterogeneous reservoirs and varying water saturations in carbonate formations (Garcia et al., 2008).

Kasza et al. (2006) reported their transition from laboratory research to practical application of emulsified acid treatments for carbonate formations, focusing on the BMB oil field, one of the largest in Poland, at an average depth of 3000 m. They faced near-wellbore damage, causing production restrictions and pressure drops, and traditional acid treatments proved ineffective due to high reservoir temperatures. They formulated a new acidizing fluid, an emulsified acid solution, consisting of a 15% hydrochloric acid solution with a corrosion inhibitor as the water phase and BMB field oil with an emulsifier as the organic phase, emulsified at a 1:1 volume ratio. This emulsion was stable at 120 °C for up to 180 min. They conducted core flow tests, which showed increased permeability in BMB reservoir rock samples. They performed a pilot treatment on the B14 well, which resulted in a twofold increase in oil production. They applied emulsified acid treatments on BMB oil and gas wells, which consistently led to significant production increases and reduced pressure differentials. They expanded this technology to neighboring oil fields with similar carbonate formations. They concluded that emulsified acidizing was a highly effective and profitable technique for enhancing acid penetration in high-temperature, low-permeability carbonate formations, and demonstrated its continued utilization in the BMB and other oil fields (Kasza et al., 2006).

1.15 Questions

Question 1-1 What is the overall objective of well-stimulation methods?

Question 1-2 What is formation damage, and why is it important to prevent it?

Question 1-3 How does clay swelling contribute to formation damage?

Question 1-4 How does drilling mud contribute to formation damage?

Question 1-5 What role do reservoir fluids play in formation damage assessment?

Question 1-6 How should well acidizing and stimulation history be evaluated?

Question 1-7 How does matrix acidizing differ from other acidizing methods?

Question 1-8 What is the significance of the wormhole phenomenon in matrix acidizing?

Question 1-9 How does acid fracturing differ from propped fracturing?

Question 1-10 How can matrix acidizing restoration of formation permeability be measured?

Question 1-11 How does the fracture network within carbonate reservoirs affect fluid transport?

Question 1-12 What factors influence acid-rock interactions during matrix acidizing?

Question 1-13 What are the benefits of using emulsified acids in carbonate reservoir acidizing?

Question 1-14 Why should the acid injection pressure be kept lower than the fracture pressure during matrix acidizing?

Question 1-15 Why are chalk and marl formations candidates for acidizing solely with hydrochloric acid (HCl)?

Question 1-16 How do the acid injection rate and reaction rate of rock influence wormhole formation?

References

Abrams, A., & Vinegar, H. (1985). Impairment mechanisms in Vicksburg tight gas sands. In *SPE Rocky Mountain Petroleum Technology Conference/Low-Permeability Reservoirs Symposium.*

Adair, P., & Smith, P. S. (1994). The miller field: Appraisal and development formation damage experiences. In *SPE Formation Damage Control Symposium.*

Adewunmi, A. A., Solling, T., Sultan, A. S., & Saikia, T. (2022). Emulsified acid systems for oil well stimulation: A review. *Journal of Petroleum Science and Engineering, 208*, 109569.

Al-Ameri, A., & Gamadi, T. (2020). Optimization of acid fracturing for a tight carbonate reservoir. *Petroleum, 6*(1), 70–79.

Al-Anazi, H. A., Walker, J. G., Pope, G. A., Sharma, M. M., & Hackney, D. F. (2003). A successful methanol treatment in a gas-condensate reservoir: Field application. In *SPE Production and Operations Symposium.*

Al-Harbi, B. G., Al-Dahlan, M., Al-Khaldi, M., Al-Harith, A., & Abadi, A. (2013). Evaluation of organic hydrofluoric acid mixtures for sandstone acidising. In *International Petroleum Technology Conference.*

Al-Harthy, S., Bustos, O. A., Samuel, M. M., & Still, J. W. (2009). Options for high-temperature well stimulation.

Al-Muhareb, M. A., Nasr-El-Din, H. A., Samuel, E., Marcinew, R. P., & Samuel, M. (2003). Acid fracturing of power water injectors: A new field application using polymer-free fluids. In *SPE European Formation Damage Conference.*

Al-Rekabi, M. A., Aktebanee, A., Al-Ghaffari, A. S., & Saleem, T. (2020). Carbonate matrix acidizing efficiency from acidizing induced skin point of view: Case study in Majnoon oilfield. In *International Petroleum Technology Conference.*

Al-Saif, A., Cochrane, J., Edmondson, H., & Youngblood, W. (1975). Analysis of pulsed-neutron decay-time logs in acidized carbonate formations. *Society of Petroleum Engineers Journal, 15*(06), 453–466.

Alhamad, L., Alrashed, A., Al Munif, E., & Miskimins, J. (2020). Organic acids for stimulation purposes: A review. *SPE Production & Operations, 35*(04), 952–978.

Ali, M., & Ziauddin, M. (2020). Carbonate acidizing: A mechanistic model for wormhole growth in linear and radial flow. *Journal of Petroleum Science and Engineering, 186*, 106776.

Ali, M. T., Ezzat, A. A., & Nasr-El-Din, H. A. (2020). A model to simulate matrix-acid stimulation for wells in dolomite reservoirs with vugs and natural fractures. *SPE Journal, 25*(02), 609–631.

Ali, M. T., & Nasr-El-Din, H. A. (2020). New insights into carbonate matrix acidizing treatments: A mathematical and experimental study. *SPE Journal, 25*(03), 1272–1284.

Ali, S. A., Kalfayan, L., & Montgomery, C. (2016). *Acid stimulation.* Monograph Series, Society of Petroleum Engineers. https://doi.org/10.2118/9781613994269.

Alvarez, J., Alarcon, X., Tellez, J., Sameer, Y., Soran, M., & Rebeen, Q. (2022). Implementing a New Foam-Acid Technology for Matrix Stimulation of Challenging Low Pressure, Naturally Fractured Carbonates Reservoirs: Case Studies, Northern Iraq. In *International Petroleum Technology Conference.*

Assem, A., Kumar, H. T., Nasr-El-Din, H. A., & De Wolf, C. (2019). Location and magnitude of formation damage due to iron precipitation during acidizing carbonate rocks. *Journal of Petroleum Science and Engineering, 179*, 337–354.

Azari, M., & Leimkuhler, J. (1990a). Completion fluid invasion simulation and permeability restoration by sodium-and potassium-based Brines. In *SPE Formation Damage Control Symposium.*

Azari, M., & Leimkuhler, J. (1990b). Formation permeability damage induced by completion brines. *Journal of Petroleum Technology, 42*(04), 486–492.

Ballard, T., & Dawe, R. (1988). Wettability alteration induced by oil-based drilling fluid. In *SPE Formation Damage Control Symposium.*

Bassir, S. M., Moghadasi, J., Daneshfar, R., Mohammadzadeh, H., & Kamranfar, P. (2017). A comprehensive review on formation damage due to fines migration. In *Fifth International Conference on Oil, Gas, Refining & Petrochemicals*, Shiraz, Iran.

Bazin, B. (2001). From matrix acidizing to acid fracturing: A laboratory evaluation of acid/rock interactions. *SPE Production & Facilities, 16*(01), 22–29.

Beletskaya, A., Ivanov, E., & Stukan, M. (2017). Reactive flow modeling at pore scale. In *SPE paper 187805 presented at the SPE Russian Petroleum Technology Conference*, Moscow, Russia, 16–18 October.

Bryant, S. L., & Buller, D. C. (1990). Formation damage from acid treatments. *SPE Production Engineering, 5*(04), 455–460.

Buijse, M., de Boer, P., Breukel, B., Klos, M., & Burgos, G. (2003). Organic acids in carbonate acidizing. In *SPE European Formation Damage Conference.*

Buijse, M., & Glasbergen, G. (2005). A semiempirical model to calculate wormhole growth in carbonate acidizing. In *Paper SPE 96892 presented at the SPE Annual Technical Conference and Exhibition*, Dallas, Texas.

Byrne, M., Rojas, E., Kandasamy, R., & Gibb, A. (2014). Fines migration in oil and gas reservoirs: Quantification and qualification through detailed study. In *SPE International Conference and Exhibition on Formation Damage Control*

Chabrelie, M.-F., Dussaud, M., Bourjas, D., & Hugout, B. (2007). Underground gas storage: Technological innovations for increased efficiency. *World Energy Council.*

Chacon, O. G., & Pournik, M. (2022). Matrix acidizing in carbonate formations. *Processes, 10*(1), 174.

Chang, F., Qu, Q., & Frenier, W. (2001). A novel self-diverting-acid developed for matrix stimulation of carbonate reservoirs In *SPE International Symposium on Oilfield Chemistry.*

Che, Y., Song, Y., Yang, H., Chen, J., Meng, C., & Zhang, J. (2023). Damage behaviour of sandstone induced by combination of dry–wet cycles and acidic environment. *Environmental Earth Sciences, 82*(1), 19.

Chen, W. L., Grabowski, R. C., & Goel, S. (2022). Clay swelling: Role of cations in stabilizing/destabilizing mechanisms. *ACS Omega, 7*(4), 3185–3191.

Crowe, C., Martin, R., & Michaelis, A. (1981). Evaluation of acid-gelling agents for use in well stimulation. *Society of Petroleum Engineers Journal, 21*(04), 415–424.

Dai, C., Zhao, F., Dai, C., & Zhao, F. (2018). Acids and additives used in acidizing. In *Oilfield chemistry* (pp. 211–235).

Dandekar, A. (2013). *Petroleum reservoir rock and fluid properties.* CRC Press.

Dankwa, O., Appah, D., Joel, O., & Asiam, E. (2016). Compatibility: a key to an efficient matrix acidizing fluid design. *Journal of Petroleum & Environmental Biotechnology, 7*, 1–5.

Daolun, L., Xuliang, L., Wenshu, Z., Jinghai, Y., & Detang, L. (2020). Automatic well test interpretation based on convolutional neural network for a radial composite reservoir. *Petroleum Exploration and Development, 47*(3), 623–631.

Dargi, M., Khamehchi, E., & Mahdavi Kalatehno, J. (2023). Optimizing acidizing design and effectiveness assessment with machine learning for predicting post-acidizing permeability. *Scientific Reports, 13*(1), 11851.

Daumer, M.-L., Picard, S., Saint-Cast, P., & Dabert, P. (2010). Technical and economical assessment of formic acid to recycle phosphorus from pig slurry by a combined acidification–precipitation process. *Journal of Hazardous Materials, 180*(1–3), 361–365.

Denney, D. (2013). Offshore water-disposal wells: Formation damage and treatment. *Journal of Petroleum Technology, 65*(11), 172–175.

Economides, M. J., & Nolte, K. G. (1989). *Reservoir stimulation* (Vol. 2). Prentice Hall Englewood Cliffs.

Elsheikh, E., Moawad, T., Alnetaifi, A., AlQuraishi, A., & Almutawea, Y. (2022). An evaluation of organic acids implementation in stimulating tight Saudi Reservoirs: A case study on Sarah sandstone formation. *Journal of King Saud University-Engineering Sciences, 34*(7), 352–358.

Fan, L., Harris, B. W., Jamaluddin, A., Kamath, J., Mott, R., Pope, G. A., Shandrygin, A., & Whitson, C. H. (2005). Understanding gas-condensate reservoirs. *Oilfield Review, 17*(4), 14–27.

Farokhi, V., & Gerami, S. (2012). A comprehensive investigation of the pseudo-skin factor for partially completed vertical wells. *Journal of Geophysics and Engineering, 9*(6), 642–654.

Fayzi, P., Mirvakili, A., Rahimpour, M., Farsi, M., & Jahanmiri, A. (2015). Experimental study of alcoholic retarded acid systems for high temperature gas wells acidizing process. *Chemical Engineering Research and Design, 93*, 576–583.

Fredd, C., & Fogler, H. S. (1998). Alternative stimulation fluids and their impact on carbonate acidizing. *SPE Journal, 3*(01), 34–41.

Fuhrberg, H. (1983). RP 5 well completion and stimulation. World Petroleum Congress.

Gao, J., Feng, P., Wang, D., Shao, S., Cui, B., & Wang, G. (2019). Large liquid volume deep penetration acidizing in Iraq Missan oilfield-3 case study. In *International Petroleum Technology Conference.*

Garcia, B., Soriano, E., Chacon, W., & Eoff, L. (2008). Novel acid-diversion technique increases production in the cantarell field, offshore Mexico. In *SPE International Conference and Exhibition on Formation Damage Control.*

Gharibi, A., & Zoveidavianpoor, M. (2015). *Hydraulic fracturing for improved oil recovery*. Department of Petroleum Engineering, Faculty of Petroleum and Renewable Energy Engineering. Universiti Teknologi, Malaysia.

Glasbergen, G., van Batenburg, D. W., Van Domelen, M. S., & Gdanski, R. D. (2005). Field validation of acidizing wormhole models. In *SPE European Formation Damage Conference.*

Gomaa, A. M., & Nasr-El-Din, H. (2010a). Rheological and core flood studies of gelled and in-situ gelled acids. In *North Africa Technical Conference and Exhibition.*

Gomaa, A. M., & Nasr-El-Din, H. A. (2010b). New insights into the viscosity of polymer-based in-situ-gelled acids. *SPE Production & Operations, 25*(03), 367–375.

Haias, H. K., & Jaf, P. T. M. Matrix acidizing of carbonate formations: A case study.

Harry, O., & Mcleod, J. (1984). Matrix acidizing. *SPE Journal, 36*(12), 2055–2069.

Hassan, A., Mahmoud, M., Al-Majed, A., Alawi, M. B., Elkatatny, S., BaTaweel, M., & Al-Nakhli, A. (2019). Gas condensate treatment: A critical review of materials, methods, field applications, and new solutions. *Journal of Petroleum Science and Engineering, 177*, 602–613.

Hassan, A., Mahmoud, M., Kamal, M. S., Hussain, S. M. S., & Patil, S. (2020). Novel treatment for mitigating condensate bank using a newly synthesized gemini surfactant. *Molecules, 25*(13), 3030.

Hemeida, A. M., & Gawish, A. A. (2008). Formation damage tests of some completion fluids. *Oil and Gas Business.*

Hendrickson, A., Thomas, R., & Economides, M. (1992). Stimulation of carbonate reservoirs. In *Developments in Petroleum Science* (Vol. 30, pp. 589–625). Elsevier.

Hoefner, M. L., & Fogler, H. S. (1987). Role of acid diffusion in matrix acidizing of carbonates. *Journal of Petroleum Technology, 39*(02), 203–208.

Houchin, L., & Hudson, L. (1986). The prediction, evaluation, and treatment of formation damage caused by organic deposition. In *SPE Formation Damage Control Symposium.*

Istchenko, C. M., & Gates, I. D. (2014). Well/wormhole model of cold heavy-oil production with sand. *SPE Journal, 19*(02), 260–269.

Jafarpour, H., Aghaei, H., Litvin, V., & Ashena, R. (2021). Experimental optimization of a recently developed matrix acid stimulation technology in heterogeneous carbonate reservoirs. *Journal of Petroleum Science and Engineering, 196*, 108100.

Kalatehno, J. M., & Khamehchi, E. (2021). A novel packer fluid for completing HP/HT oil and gas wells. *Journal of Petroleum Science and Engineering, 203*, 108538.

Kalatehno, J. M., Khamehchi, E., Abbasi, A., & Khaleghi, M. R. (2023). A novel approach to determining appropriate additive concentrations for stimulation of gas carbonate reservoirs. *Results in Engineering*, 101440.

Kalhor Mohammadi, M., Riahi, S., & Boek, E. S. (2023). An insight review on formation damage induced by drilling fluids. *Reviews in Chemical Engineering, 39*(3), 387–415.

Kargarpour, M. A. (2020). Carbonate reservoir characterization: An integrated approach. *Journal of Petroleum Exploration and Production Technology, 10*(7), 2655–2667.

Kasi, N., Justine, Y., Madeleine, N., & Théophile, N.-M. (2022). Remediation of oil production by matrix acidizing method on an oil well. *Iraqi Journal of Chemical and Petroleum Engineering, 23*(4), 1–6.

Kasza, P., Dziadkiewicz, M., & Czupski, M. (2006). From laboratory research to successful practice: A case study of carbonate formation emulsified acid treatments. In *SPE International Conference and Exhibition on Formation Damage Control.*

Keihani Kamal, M., Mahdavi Kalatehno, J., Daneshfar, P., & Yousefmarzi, F. (2024, 2024/01/17). A comprehensive analysis of carbonate matrix acidizing using viscoelastic diverting acid system in a gas field. *Scientific Reports, 14*(1), 1499. https://doi.org/10.1038/s41598-024-52104-5

Khuzin, R. A., & Khizhniak, G. P. (2019). Laboratory research of acid concentration and injection rate on worm holing process under reservoir conditions. *Perm Journal of Petroleum and Mining Engineering, 19*(4), 356–372.

Krueger, R. F. (1986). An overview of formation damage and well productivity in oilfield operations. *Journal of Petroleum Technology, 38*(02), 131–152.

Leong, V. H., & Ben Mahmud, H. (2019). A preliminary screening and characterization of suitable acids for sandstone matrix acidizing technique: A comprehensive review. *Journal of Petroleum Exploration and Production Technology, 9*(1), 753–778.

Letichevskiy, A., Nikitin, A., Parfenov, A., Makarenko, V., Lavrov, I., Rukan, G., Ovsyannikov, D., Nuriakhmetov, R., & Gromovenko, A. (2017). Foam acid treatment-The key to stimulation of carbonate reservoirs in depleted oil fields of the Samara region. In *SPE Russian petroleum technology conference.*

Li, L., Nasr-El-Din, H. A., Chang, F. F., & Lindvig, T. (2008). Reaction of simple organic acids and chelating agents with calcite. In *International Petroleum Technology Conference.*

Lian, J., Wang, P., Wang, H., Guo, Y., Xu, Y., Ye, F., & Yang, Y. (2021). Experimental study of one-step overall transportation of composite bucket foundation for offshore wind turbine under the coupled dynamic action of wave and current. *Journal of Renewable and Sustainable Energy, 13*(3), 035701.

Liu, P., Li, J., Sun, S., Yao, J., & Zhang, K. (2021). Numerical investigation of carbonate acidizing with gelled acid using a coupled thermal–hydrologic–chemical model. *International Journal of Thermal Sciences, 160*, 106700.

Liu, P., Salem, A. M., Pochetnyy, V., Alghannam, A., Bosalhah, A., Almulla, S., Duggirala, V. S., Ayyavoo, M. M., Al-Bader, H., & Al-Obaidi, A. (2023). First successful openhole multistage completion and acid fracturing treatment to unlock the production potential in an HP/HT, high-H2S exploratory well in North Kuwait. In *Abu Dhabi International Petroleum Exhibition and Conference.*

Lungwitz, B., Fredd, C., Brady, M., Miller, M., Ali, S., & Hughes, K. (2007). Diversion and clean up studies of viscoelastic surfactant-based self-diverting acid. *SPE Production & Operations, 22*(01), 121–127.

Mahmoud, M., Aljawad, M. S., Kamal, M. S., & Hussain, S. M. S. (2020). Sandstone acidizing using a new retarded acid system based on gemini surfactants. *Journal of Petroleum Science and Engineering, 194*, 107459.

McDuff, D. R., Shuchart, C. E., Jackson, S. K., Postl, D., & Brown, J. S. (2010). Understanding wormholes in carbonates: Unprecedented experimental scale and 3-D visualization. In *SPE Annual Technical Conference and Exhibition.*

McLeod, H. O. (1984). Matrix acidizing. *Journal of Petroleum Technology, 36*(12), 2055–2069.

McPhee, C., Reed, J., & Zubizarreta, I. (2015). *Core analysis: A best practice guide.* Elsevier.

Morrow, N. (1998). Evaluation of reservoir wettability and its effect on oil recovery. *Journal of Petroleum Science and Engineering,* 1–301.

Mumallah, N. (1991). Factors influencing the reaction rate of hydrochloric acid and carbonate rock. In *SPE International Symposium on Oilfield Chemistry.*

Nasr-El-Din, H., Driweesh, S., & Muntasheri, G. (2003). Field application of HCl-formic acid system to acid fracture deep gas wells completed with super Cr-13 tubing in Saudi Arabia. In *SPE International Improved Oil Recovery Conference in Asia Pacific.*

Nasr-El-din, H., Driweesh, S., & Muntasherl, G. (2004). Field application of an HCl/formic acid system to acid fracture deep gas wells. *Journal of Petroleum Technology, 56*(3), 59–60.

Nasr-El-Din, H., Lynn, J., & Taylor, K. (2001). Lab testing and field application of a large-scale acetic acid-based treatment in a newly developed carbonate reservoir. In *SPE International Symposium on Oilfield Chemistry.*

Navarrete, R., Holms, B., McConnell, S., & Linton, D. (2000). Laboratory, theoretical, and field studies of emulsified acid treatments in high-temperature carbonate formations. *SPE Production & Facilities, 15*(02), 96–106.

Noshi, C. I. (2015). *A novel in-situ generated acid system on carbonate mineral dissolution in sandstone reservoirs*

Palharini Schwalbert, M., Aljawad, M. S., Hill, A. D., & Zhu, D. (2020). Decision criterion for acid-stimulation method in carbonate reservoirs: Matrix acidizing or acid fracturing? *SPE Journal, 25*(05), 2296–2318.

Panga, M. K., Ziauddin, M., & Balakotaiah, V. (2005). Two-scale continuum model for simulation of wormholes in carbonate acidization. *AIChE Journal, 51*(12), 3231–3248.

Rabbani, E., Davarpanah, A., & Memariani, M. (2018). An experimental study of acidizing operation performances on the wellbore productivity index enhancement. *Journal of Petroleum Exploration and Production Technology, 8*(4), 1243–1253.

Radwan, A. E., Wood, D. A., Abudeif, A., Attia, M., Mahmoud, M., Kassem, A. A., & Kania, M. (2022). Reservoir formation damage; reasons and mitigation: A case study of the Cambrian-Ordovician Nubian 'C'Sandstone Gas and Oil Reservoir from the Gulf of Suez Rift Basin. *Arabian Journal for Science and Engineering, 47*(9), 11279–11296.

Rahaman, F., Paul, N., Banerjee, A., De, S., Ray, S., & Usmani, A. (2016). The Finslerian wormhole models. *The European Physical Journal C, 76*, 1–9.

Schechter, R. S. (1992). *Oil well stimulation.*

Shafiq, M. U., Chong, Y. J., Mahmud, H. K. B., Hossain, M. M., Rezaee, R., & Testamanti, N. (2019). Application of emulsified acids on sandstone formation at elevated temperature conditions: An experimental study. *Journal of Petroleum Exploration and Production Technology, 9*, 1323–1329.

Shajari, M., & Malhani, H. (2023). Stimulation of low pressure carbonates reservoirs with a triphase emulsified acid system "3PEA" of Southern Iran-case studies. *Improved Oil and Gas Recovery, 7*.

Smith, C. F., & Hendrickson, A. R. (1965). Hydrofluoric acid stimulation of sandstone reservoirs. *Journal of Petroleum Technology, 17*(02), 215–222. https://doi.org/10.2118/980-pa

Sousa, A. M., Pereira, M. J., & Matos, H. A. (2022). Oil-in-water and water-in-oil emulsions formation and demulsification. *Journal of Petroleum Science and Engineering, 210*, 110041.

Sui, Y., Cao, G., Guo, T., Li, Z., Bai, Y., Li, D., & Zhang, Z. (2022). Development of gelled acid system in high-temperature carbonate reservoirs. *Journal of Petroleum Science and Engineering, 216*, 110836.

Tannich, J. (1975). Liquid removal from hydraulically fractured gas wells. *Journal of Petroleum Technology, 27*(11), 1309–1317.

Taylor, K., & Nasr-El-Din, H. (2003). Laboratory evaluation of in-situ gelled acids for carbonate reservoirs. *SPE Journal, 8*(04), 426–434.

Taylor, K. C., Nasr-El-Din, H. A., & Mehta, S. (2006). Anomalous acid reaction rates in carbonate reservoir rocks. *SPE Journal, 11*(04), 488–496.

Wang, C., Zhao, L., He, B., Zhang, X., & Mmbuji, A. O. Comparative analysis of matrix retarded acidizing methods for tight carbonate reservoirs: Gelled acid, foamed acid and micro-emulsified acid. *Frontiers in Earth Science, 2142*.

Wang, Q. (2021). *Fluid chemistry, drilling and completion*. Gulf Professional Publishing.

Wang, R., Lv, N., Li, C., Cai, G., Pan, X., Li, Y., & Zhu, G. (2021). Novel strategy for enhancing acetic and formic acids generation in acidogenesis of anaerobic digestion via targeted adjusting environmental niches. *Water Research, 193*, 116896.

Wang, Y., Hill, A. D., & Schechter, R. S. (1993). The optimum injection rate for matrix acidizing of carbonate formations. In *SPE Annual Technical Conference and Exhibition*.

Xu, C., Kang, Y., You, Z., & Chen, M. (2016). Review on formation damage mechanisms and processes in shale gas reservoir: Known and to be known. *Journal of Natural Gas Science and Engineering, 36*, 1208–1219.

Yan, Y., Guan, Z., Xu, Y., Yan, W., & Wang, H. (2019). Numerical investigation of perforation to cementing interface damage area. *Journal of Petroleum Science and Engineering, 179*, 257–265.

Yan, Y., Guan, Z., Yan, W., & Wang, H. (2020). Mechanical response and damage mechanism of cement sheath during perforation in oil and gas well. *Journal of Petroleum Science and Engineering, 188*, 106924.

Yew, C. H., & Weng, X. (2014). *Mechanics of hydraulic fracturing*. Gulf Professional Publishing.

You, J., & Lee, K. J. (2021). Analyzing the dynamics of mineral dissolution during acid fracturing by pore-scale modeling of acid-rock interaction. *SPE Journal, 26*(02), 639–652.

Zhao, X., Qiu, Z., Sun, B., Liu, S., Xing, X., & Wang, M. (2019). Formation damage mechanisms associated with drilling and completion fluids for deep water reservoirs. *Journal of Petroleum Science and Engineering, 173*, 112–121.

Zhou, F., Liu, Y., Yang, X., Lian, S., Zhang, F., He, J., Xiong, C., Liu, X., Shi, Y., & Jia, X. (2009). A novel diverting acid stimulation treatment technique for carbonate reservoirs in China. In *Asia Pacific Oil and Gas Conference & Exhibition*.

Chapter 2
Estimating the Formation Fracture Pressure Through Geomechanical Modeling

2.1 Introduction

The geomechanical model plays a vital role in the exploration, production, and development of oil and gas reservoirs. It involves continuous numerical representation of changes in rock mechanical properties, ground stress, and pore pressure. This model is indispensable for tasks such as hydraulic fracture design, ensuring well stability, and facilitating drilling operations. By using a geomechanical model, significant cost reduction and technical improvements in related processes can be achieved. To create a geomechanical model, various data such as porosity, density, and acoustic measurements are typically utilized. Calibration of the model is done through core and downhole tests to ensure accuracy. One-dimensional geomechanical models provide quantitative insights into in-situ stresses and physical properties within a field or reservoir. These models are constructed using data from well surveys and can be used to assess aspects like wellbore stability, mud management, sand production, well-drilling safety, hydraulic fracture initiation, and drilling path planning. In the oil and gas industry, geomechanics is of paramount importance. It is a discipline that investigates and analyzes the behavior of the Earth's subsurface under various stresses. These stresses can be natural or induced by drilling operations. Therefore, geomechanics addresses a wide range of issues including wellbore stability, hydraulic fracture performance, sand production, and well casing integrity, among others.

Reservoir geomechanics, a multidisciplinary field encompassing rock mechanics, structural geology, and petroleum engineering, aims to identify and resolve challenges encountered during the exploration and production of oil and gas reservoirs. Geomechanical models provide numerical representations of stress distribution and rock properties across a stratigraphic column within a field. Additionally, rock physics relationships are employed to correlate mechanical properties with reservoir petrophysical and geophysical data. Many oil fields grapple with geomechanical issues such as overpressure, well instability, reservoir compaction, casing failures,

© The Author(s), under exclusive license to Springer Nature Switzerland AG 2024

E. Khamehchi et al., *Applied Matrix Acidizing of Carbonate Reservoir*, Petroleum Engineering, https://doi.org/10.1007/978-3-031-58281-3_2

sand production, subsidence, and fault activity. Implementing geomechanical models early in the reservoir's production life can significantly mitigate these risks and streamline field development. Building a comprehensive geomechanical model involves addressing uncertainties in various model components. The sooner this model is developed during reservoir production, the quicker it can help prevent and resolve problems, ultimately contributing to the optimal field development. These models find applications in drilling, well management, formation assessment, well completion, stimulation, and production engineering, among other areas.

Necessary Parameters for Creating a Geomechanical Model (Liu & Znidarčić, 1991).

The first category encompasses parameters related to the formation, which includes the elastic constants of the rock, resistance parameters, physical properties Biot's coefficient (α), and pore fluid pressure $\left(P_p\right)$.

The second group of parameters pertains to regional stress, encompassing values of vertical stresses (σ_V), maximum (σ_H), and minimum horizontal stresses (σ_H), pore fluid pressure, and their respective directions.

2.2 Elasticity Parameters

Most materials possess the ability to resist deformation caused by external forces and can recover from it. This inherent capability is referred to as elasticity and forms the foundation of various aspects of rock mechanics. The simplest type of response occurs when there is a linear relationship between the applied external forces and the resulting deformations. When these force changes remain relatively small, the response is (almost always) linear. Therefore, the theory of linear elasticity underpins and informs all discussions related to tension (Martínez-Martínez et al., 2012).

2.2.1 Stress

A crucial element in developing a comprehensive geomechanical model involves understanding the current state of stress. Well failures occur when the concentrated stress around the well exceeds the rock's resistance. Compression stress is omnipresent in subsurface strata and varies in magnitude with depth, pore pressure, and the active geological processes occurring on different spatial and temporal scales. In simpler terms, stress represents the force applied over a specific area. More precisely, stress is a tensor that characterizes the forces acting on all surfaces passing through a given point.

To illustrate, consider the scenario depicted in Fig. 2.1. A weight is positioned on top of a column, creating a force exerted on the column, which, in turn, reacts with an equal yet opposite force. The column itself is supported by the ground, meaning that the force applied at the top of the column affects any section of the column. Let's

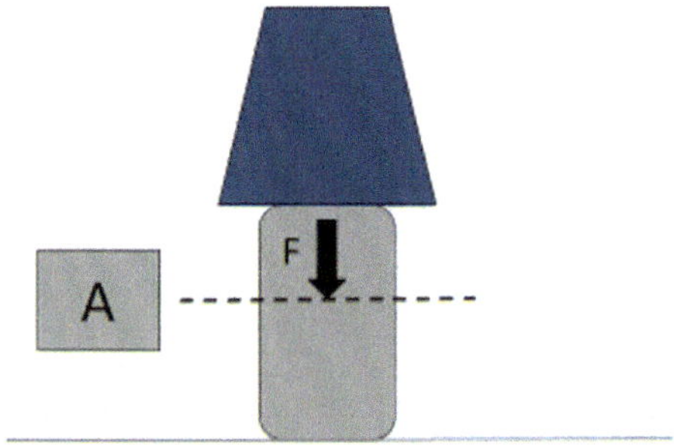

Fig. 2.1 Image of forces and stress

denote the cross-sectional area at point 'a' as 'A'. If we represent the force acting on this cross-section as 'F', the stress 'σ' at cross-section 'A' can be calculated using Eq. 2.1 (Zoback, 2010):

$$\sigma = \frac{F}{A} \tag{2.1}$$

The SI unit for stress is pascal (Pascal = N/m^2).

2.2.2 *Strain*

Strain refers to the alteration in the size or shape of an object resulting from the application of an external force. When the relative positions of particles within a sample change in such a way that these new positions cannot be achieved through a simple translation or rotation of the sample, the sample is considered to be strained. Figure 2.2 illustrates an example of a strained specimen. Equation 2.2 represents the most basic relation associated with strain.

$$\varepsilon = \frac{L - L'}{L} = -\frac{\Delta L}{L} \tag{2.2}$$

ε is the strain, L is the initial length and L′ is the final length of the object (Zoback, 2010).

Example 2.1 A force of 160 N is applied to a cube as shown, and the length of each side of the cube is 5 cm. How much stress does the cube get when the force is applied?

Fig. 2.2 Deformation

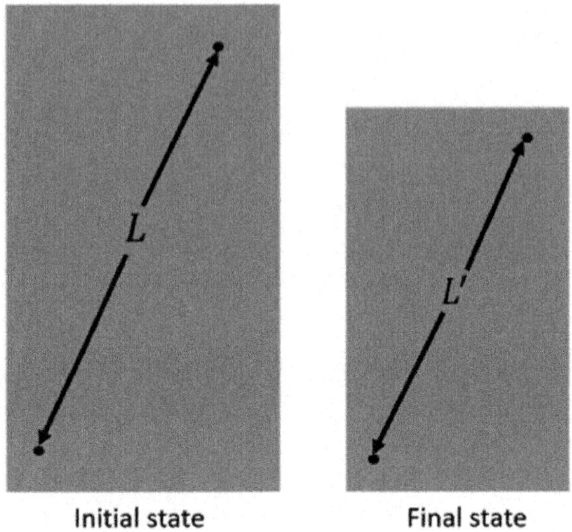

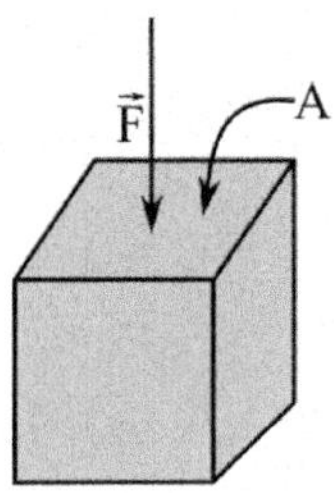

Answer:

$$F = 160\ \text{N},\ A = 25\ \text{cm}^2 = 0.025\ \text{m}^2$$

$$\sigma = \frac{F}{A} = \frac{160\ \text{N}}{0.025\ \text{m}^2} = 6400\frac{\text{N}}{\text{m}^2} = 6.4\ \text{kpa}$$

B: If final length of the object achieved to 3 cm, calculate strain.

$$\varepsilon = \frac{L - L'}{L} = -\frac{\Delta L}{L} = -\frac{3\ \text{cm} - 5\ \text{cm}}{3\ \text{cm}} = 0.667$$

2.2.3 *Stress and Strain Relationship*

When a piece of rock is subjected to sufficiently high pressures, it can experience a type of failure, leading to a change in its shape and a potential for disintegration. This

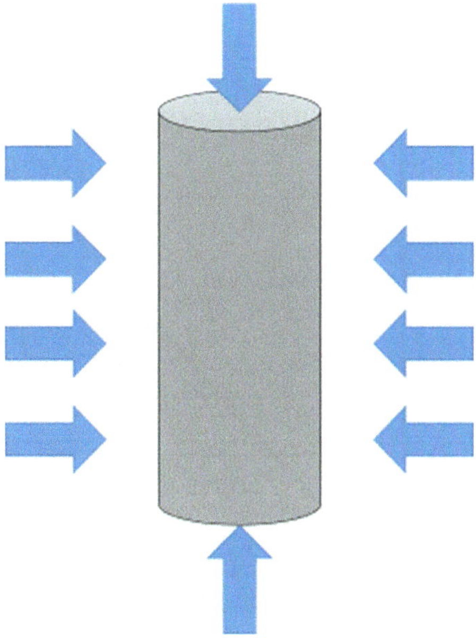

Fig. 2.3 A standard test specimen used for uniaxial or triaxial testing

situation often results in a decrease in the rock's load-carrying capacity. The stress level at which rock typically breaks is referred to as its rock strength. In Fig. 2.3, you can see a cylinder with a length-to-diameter ratio of 1:2. Two pistons apply axial stress to the cylinder's end faces, and the cylinder is surrounded by an oil bath that creates varying stress levels in the surrounding material. It's generally assumed that the stress distribution is uniform within the sample. When there is no confining stress, it's known as a uniaxial stress test (also called an Unconfined Compression Test). When the test is conducted with non-zero confining pressure, it's referred to as a triaxial test.

Figure 2.4 presented in illustrates a standard outcome of a uniaxial test, where the applied axial stress (represented as σ_z) is graphed against the axial strain (ε_z) of the sample.

- **Elastic Zone**: This is where the rock deforms elastically. If the stress is removed, the sample returns to its original state.
- **Yield Point**: Beyond this point, any changes in the sample will be permanent. Even with a stress reduction, the sample will not return to its original state.
- **Uniaxial Compressive Strength**: This is the maximum applied stress.
- **Ductile Zone**: In this zone, the sample undergoes permanent deformation without losing its load-bearing capacity.
- **Brittle Zone**: This is the region where the sample's ability to bear stress decreases rapidly with increased deformation (Zoback, 2010).

According to the information provided, rock fracture occurs when stress exceeds a certain threshold, while the rock remains intact as long as the stress is below this

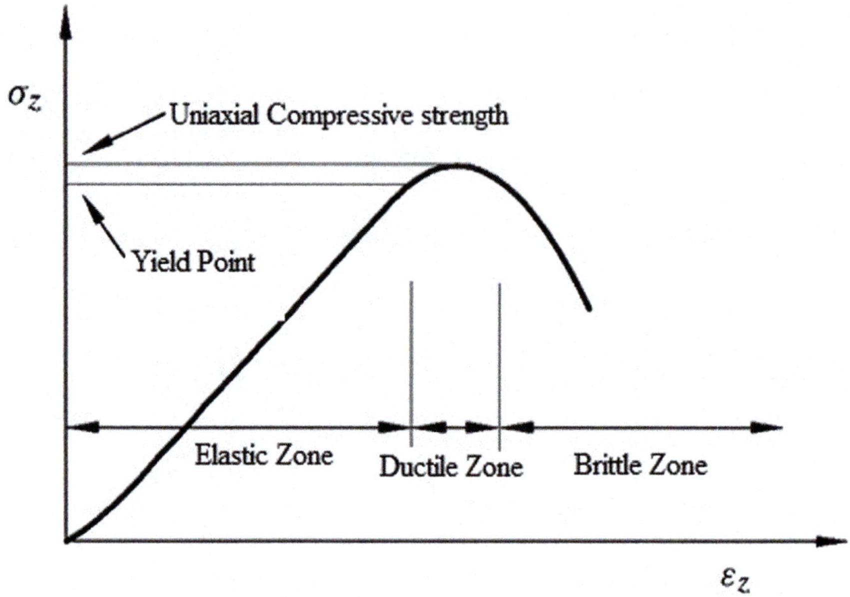

Fig. 2.4 Stress versus deformation in a uniaxial compression test (in practice, the ductile area may be very small)

threshold. Two common types of rock failure are tensile and shear failure. Tensile failure happens when the effective tensile stress on a surface of the specimen surpasses a critical limit, denoted as tensile strength (T_0) and measured in stress units. As illustrated in Fig. 2.5, a specimen undergoing tensile failure typically failures along specific planes. Shear failure (τ) occurs when the shear stress along certain surfaces of the sample is sufficiently high. Ultimately, this leads to the development of a fault zone along the fault plane, with the two sides of the plane moving relative to each other through a frictional process, as depicted in Fig. 2.6.

To describe failure behavior, the Mohr–Coulomb criterion is employed, which is a widely accepted and general criterion (Fig. 2.7). This criterion is founded on the premise that τ (shear stress) is a linear function of the normal stress (σ) (as expressed

Fig. 2.5 Tensile failure

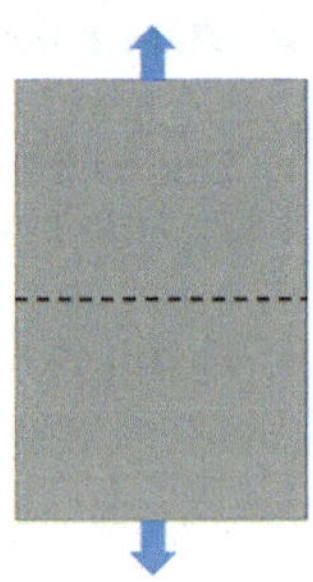

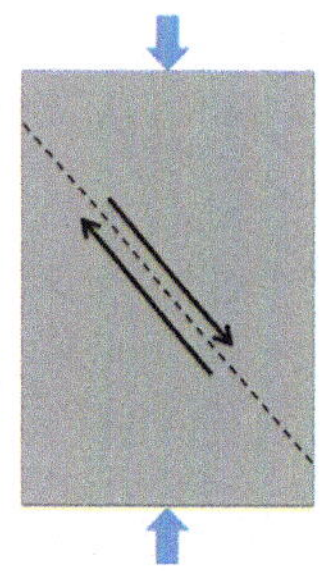

Fig. 2.6 Shear failure

in Eq. 2.3). When Mohr's circle intersects the plotted line, it indicates that the sample will undergo shear failure due to the normal stress.

$$\tau = S_o + (\tan\varphi)\tau \quad (2.3)$$

S_o is the inherent shear strength, which is also called Cohesion. The angle ϕ defined in the figure is called the coefficient of internal friction (Friction angle) and is related to the coefficient of internal friction (Eq. 2.4) (Bai & Wierzbicki, 2010; Hoek, 1990; Labuz & Zang, 2012).

$$\tan\varphi = \mu \quad (2.4)$$

Example 2.2 A soil sample is subjected to a normal stress of 100 kPa. The inherent shear strength (cohesion) of the soil is 50 kPa and the coefficient of internal friction

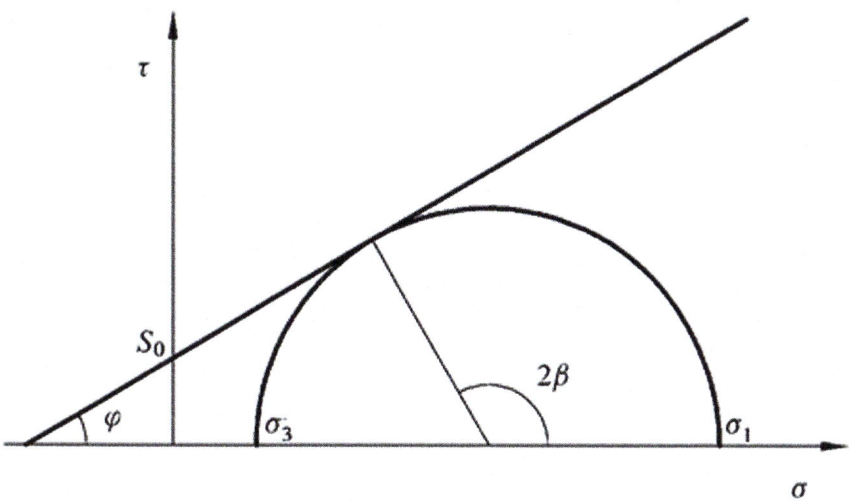

Fig. 2.7 Mohr coulo criterion

(friction angle) is 30°. Determine the shear stress acting on the soil sample using the Mohr–Coulomb criterion.

Answer:
Given:
Normal stress (σ) = 100 kPa.
Inherent shear strength (Cohesion, S_o) = 50 kPa.
Coefficient of internal friction (Friction angle, φ) = 30°.
Using Eq. 2.3:

$$\tau = S_o + (\tan\varphi)\sigma$$

Substituting the given values:

$$\tau = 50\text{ kPa} + \tan(30^\circ)(100\text{ kPa})$$

Calculating the tangent of 30°:

$$\tan(30^\circ) \approx 0.5774$$

$$\tau \approx 50\text{ kPa} + (0.5774)(100\text{ kPa})$$

$$\tau \approx 50\text{ kPa} + 57.74\text{ kPa}$$

$$\tau \approx 107.74\text{ kPa}$$

Therefore, the shear stress acting on the soil sample is approximately 107.74 kPa according to the Mohr–Coulomb criterion.

2.3 Uniaxial Compressive Strength

The uniaxial compressive strength test is the most common laboratory test for mechanical studies of virgin rocks, which, despite its relatively simple appearance, is very difficult to perform accurately. This test is performed as a basic test in most engineering projects and it rarely happens that it is not needed in a uniaxial compressive strength project. Although this test is mainly known as an index for comparing rocks, it has other wide applications in solving practical problems of rock mechanics. As can be seen in the next figure, with the help of a uniaxial compressive strength test, the stress–strain curve of the rock can be drawn and the yield strength of the rock and the ultimate strength of the rock can be obtained from it (Chau & Wong, 1996; Kahraman, 2001; Teymen & Mengüç, 2020) (Fig. 2.8).

Fig. 2.8 Uniaxial compressive strength test

2.4 Pore Pressure

Pore pressure is a significant parameter in the geomechanical evaluation of oil and gas reservoirs. It plays a crucial role in well drilling, mud weight design, and wellbore integrity. Pore pressure can vary from being equal to hydrostatic pressure (the height of the fluid column) to approaching lithostatic pressure in different parts of a field. Estimating pore pressure is essential for risk reduction and improved well design.

There are various methods for estimating pore pressure, including:

1. Penetration pressure based on rock resistance.
2. Penetration pressure, is determined by measuring the time sound waves travel.
3. Pore pressure estimation based on porosity.
4. Penetration pressure in shale formations.
5. Pore pressure evaluation using hydraulic methods.
6. Estimating pore pressure based on the weight of drilling mud.

Pore pressure is a critical factor in rock mechanics, particularly in porous, fluid-filled rock systems. The presence of pore fluid redistributes some of the applied stresses within the system, reducing the load on the rock matrix. Terzaghi's concept of effective stress, introduced experimentally in 1923 in soil mechanics, is essential in understanding this equation.

$$\sigma' = \sigma - P_p \tag{2.5}$$

In 1957, Biot and Willis introduced Eqs. 2.6 and 2.7 to refine the concept of effective stress in porous rocks that contain fluids. Equation 2.7 demonstrates that Biot's coefficient, denoted as α, represents the ratio of the bulk modulus under constant pore pressure conditions to the bulk modulus under constant confining pressure conditions.

$$\sigma' = \sigma - \alpha P_p \tag{2.6}$$

$$\alpha = \frac{K_{fr}}{K_{bp}} \tag{2.7}$$

α, often referred to as Biot's porosity constant, is a parameter that characterizes how efficiently fluid pressure counteracts total stress within a rock. Its magnitude is influenced by the pore structure and physical properties of the solid components in the rock system. A higher α value indicates that the rock is more easily compressible. In cases where the compressibility of dry rock greatly exceeds that of the solid rock grains, which is typical for sedimentary reservoir rocks, α approaches 1. Values of α ranging from 0.5 to 1 have been measured for various reservoir rocks, with the range of 0.9–1.0 being commonly used in Special Core Analysis Laboratory (SCAL) testing. While α can be challenging to measure experimentally, rock and matrix bulk moduli or compressibility can be derived from DSI plot data (Salemi et al., 2018; Zhang, 2011).

2.5 Estimation of the Magnitude of the Main Stresses in Place

The estimation of the magnitude of the principal stresses in situ is a critical aspect of geomechanical studies.

In-situ stresses refer to the natural stresses within the Earth's crust, encompassing both horizontal and vertical components. These stresses have various origins:

1. Gravitational Stresses: These stresses result from the weight of overlying strata.
2. Tectonic Stresses of the Present Era: These stresses are associated with contemporary tectonic forces. A prominent example is the movement and interaction of the Arabian plate with the Eurasian plate in the Persian Gulf region.
3. Residual Stresses: Rock residual stresses are linked to past tectonic and gravitational forces.

In the coordinate system relative to the Earth's surface, these stresses are categorized as vertical stress, maximum horizontal stress, and minimum horizontal stress. The magnitude and direction of these principal stresses within a borehole are crucial factors influencing wellbore stability, drilling path design, production, and geomechanical behavior.

Typically, in-situ stresses are represented as tensors with components σ. However, in oil field studies, since wells are generally drilled vertically, stress analysis focuses on three primary components: the vertical component (σ_V), as well as the minimum and maximum horizontal components (σ_h, σ_H).

As a result, various direct and indirect methods, including experimental relationships, have been developed to determine the magnitude of minimum and maximum horizontal stress in oil reservoirs and at significant depths. Nevertheless, accurately determining the principal horizontal stresses, especially the maximum horizontal stress, remains a fundamental challenge in geomechanical studies.

2.5.1 *Vertical Tension* (σ_V)

One of the primary stresses is the vertical stress, also known as the overburden stress, which results from the weight of the overlying layers. The other two significant stresses are the maximum and minimum horizontal stresses. The vertical stress can be determined through a density diagram. In cases where data is unavailable, density can be estimated using established empirical relationships.

$$\sigma_V = \int_0^Z \rho(z) g dz \cong \overline{\rho} g z \tag{2.8}$$

Here, $\overline{\rho}$ represents the average density, g denotes gravitational acceleration, and $\rho(z)$ represents rock density as a function of depth.

The challenge in this context lies in the fact that the density profile is typically measured within reservoir sections. In such cases, it becomes necessary to estimate density values above the reservoir layer up to ground level through extrapolation methods. Consequently, rock density near the Earth's surface is approximated as 1.8. By employing a logarithmic function fitting, a relationship can be derived to compute the density in areas where specific values are not available.

Example 2.3 Calculate vertical stress in reservoir at depth of 3000 m. Assume all top layer are dry calcite.

Answer:

$$\begin{aligned}\sigma_V &= \int_0^Z \rho(z) g dz \cong \overline{\rho} g z \\ &= 2711\ \frac{\text{kg}}{\text{m}^3} \times 9.8\ \frac{\text{N}}{\text{kg}} \times 3000\ \text{m} \\ &= 79703.4\ \frac{\text{kN}}{\text{m}^2} = 79703.4\ \text{kpa}\end{aligned}$$

Table 2.1 The most crucial horizontal stress relationships are at a minimum

Correlation	Procedure
$S_{hmin} = 0.3(\sigma_v - P_p) + P_p$	Hubbert and Willis (1957)
$S_{hmin} = \frac{\sigma_h}{\sigma_v}(\sigma_v - P_p) + P_p$	Matthews (1967)
$S_{hmin} = \left(\frac{\vartheta}{1-\vartheta}\right)(\sigma_v - P_p) + P_p$	Eaton (1969)
$S_{hmin} = 0.197H^{1.145} + 0.46(P_p - P_h) for H < 1150 ft$ $S_{hmin} = 1.167H - 4596 + 0.46(P_p - P_h) for H > 1150 ft$	Breckels and Van Eekelen (1982)
$S_{hmin} = \frac{\sigma_h - P_p}{\sigma_v - P_p} = \left[(\mu^2 + 1)^{0.5} + \mu\right]^{-2}$	Byerlee and Zoback (1975)
$S_{hmin} = (1 - \varphi)(\sigma_v - P_p) + P_p$	Holbrook (1990)

2.5.2 *The Minimum Horizontal Stress* (σ_h)

The minimum horizontal stress, denoted as σ_h, plays a crucial role in solving various geomechanical problems. Calculating this parameter is essential for optimizing well completion operations, designing drilling processes, and expanding hydraulic fracturing activities. There are several methods available for determining the minimum horizontal stress, which are outlined below:

- Hydraulic fracturing test
- Leak-off test
- Mini Fracture Test

Given the substantial costs associated with direct testing, the utilization of empirical relationships has gained prominence in recent times. To estimate the minimum horizontal stress, numerous significant empirical relationships are provided in Table 2.1.

2.5.3 *Maximum Horizontal Stress* (σ_H)

Maximum horizontal stress, denoted as (σ_H), is a critical parameter in many geomechanical problems, although accurately estimating it can be quite challenging. Understanding the maximum horizontal stress is essential for addressing various geomechanical issues, including those related to well stability. This encompasses tasks such as designing the well path, determining the optimal location for casing, and calculating the appropriate weight of drilling mud.

The most significant methods for estimating the maximum horizontal stress include:

- Hydraulic fracture test
- Utilizing experimental relationships

These approaches are instrumental in providing valuable insights into the maximum horizontal stress.

2.6 Determining the Direction of the Main Stresses in Place

Numerous studies have been conducted worldwide on stress direction, with the results compiled in the 'Global Stress Map' project, which is regularly updated to accommodate new information. While this map contains data from various regions, including the Middle East and especially Iran, it is important to note that information regarding tension sources in oil fields in these areas, which should ideally be determined through spill analysis and wellbore failure assessments, remains limited. There are several methods for determining the direction of horizontal stresses, each suited to specific situations. These methods include earthquake center mechanisms, the use of castings or tensile failure on wellbore walls, hydraulic techniques, jacking methods, and geological indicators like fault slips. Each technique provides insights into the stress field at different depths. For instance, earthquake center mechanisms offer data on in situ stress fields at depths ranging from 5 to 40 km, while hydraulic methods are typically applicable to depths less than 1 km, such as engineering projects. Geological indicators like fault slip analysis reveal surface stress data. However, one of the most commonly used methods for determining in situ stresses, particularly in hydrocarbon fields, is the study of castings and tensile failure created during drilling operations. Understanding the current tectonic stress is fundamental in oil exploration and production, with key applications including well stability, reservoir production and fluid flow patterns, natural fractured reservoir fluid flow, hydraulic fracturing stimulation, reactivation of faults, and stress orientation in smaller and regional scales. Accurate identification of small-scale fractures around wells is challenging, as most methods such as well survey logs, well tests, and mud hole data lack the necessary resolution. Image logs, on the other hand, offer high-resolution well wall images down to the millimeter level, providing vital information about well wall characteristics such as electrical resistance and sound contrast. These logs enable reservoir geologists and petroleum engineers to detect small-scale phenomena, crucial for well completion and oil field development planning, including diversion well routing. The primary application of image logs is to identify fracture types and their attributes, including type, orientation, and aperture, in oil and gas wells. Image logs also find diverse uses, including sedimentological studies for reservoir rock texture determination, formation factor and cementation coefficient calculation, secondary porosity assessment, in situ stress determination, and heterogeneity analysis.

2.7 Image Logs (FMI)

Image logs are virtual cylindrical representations of the wellbore wall, providing high-resolution views capable of revealing subtle wall features. Any flat-shaped phenomena, such as layers and fractures that intersect the wellbore at non-vertical angles, appear as ovals within the cylindrical image. If the cylinder is cut along its axis and unfurled, it assumes the same shape as seen in the image log, resembling a sine wave.

The Formation MicroScanner (FMI) is an electrical imaging tool comprising two pairs of vertical caliper arms, each linked to a set of pads and flaps at their ends, housing multiple resistance sensors. These sensors record data used to generate an "image" based on variations in wellbore wall resistance. In addition to resistivity imaging, the FMI tool collects information about wellbore cross-sectional shape and geometry. The FMI image serves as a valuable tool for interpreting fractures resulting from drilling, which align with the S_{Hmax} (maximum horizontal stress) direction, as well as for interpreting Symmetric Wellbore Collapses perpendicular to S_{Hmax}.

Moreover, FMI logs are instrumental in interpreting the geometry of natural fractures, which appear as sinusoidal patterns on the logs and may exhibit electrical resistance or conductivity characteristics (Fig. 2.9).

2.8 Elastic Moduli

Linear elasticity theory addresses situations where there exists a linear correlation between the applied stress and strain. Although many rocks exhibit nonlinear behavior under high stresses, their response to low stresses is often described by linear relationships. Let's consider an example involving a specimen with a length 'L' and a cross-sectional area '*A*' (represented as $A = D^2$ in Fig. 3.3) When a force '*F*' is applied to its end surfaces, the sample's length contracts to L'. Subsequently, the stress applied can be calculated as $\sigma_x = \frac{F}{A}$, and the corresponding length change can be determined using the equation $\varepsilon_x = \frac{(L-L')}{L}$. In the case of linear behavior, there exists a linear connection between ε_x and σ_x.

$$\varepsilon_x = \frac{1}{E}\sigma_x \tag{2.9}$$

Equation 2.9 is commonly referred to as Hooke's law, with the coefficient '*E*' known as Young's modulus or simply '*E*' modulus. Young's modulus falls within a category of coefficients known as elastic moduli. It quantifies the stiffness of a material, specifically its resistance to compression under uniaxial stress.

Another effect of the applied stress σ_x (as depicted in Fig. 2.10 is the change in the width '*D*' of the sample. This lateral expansion is quantified by $\varepsilon_y = \varepsilon_z = \frac{(D-D')}{D}$.

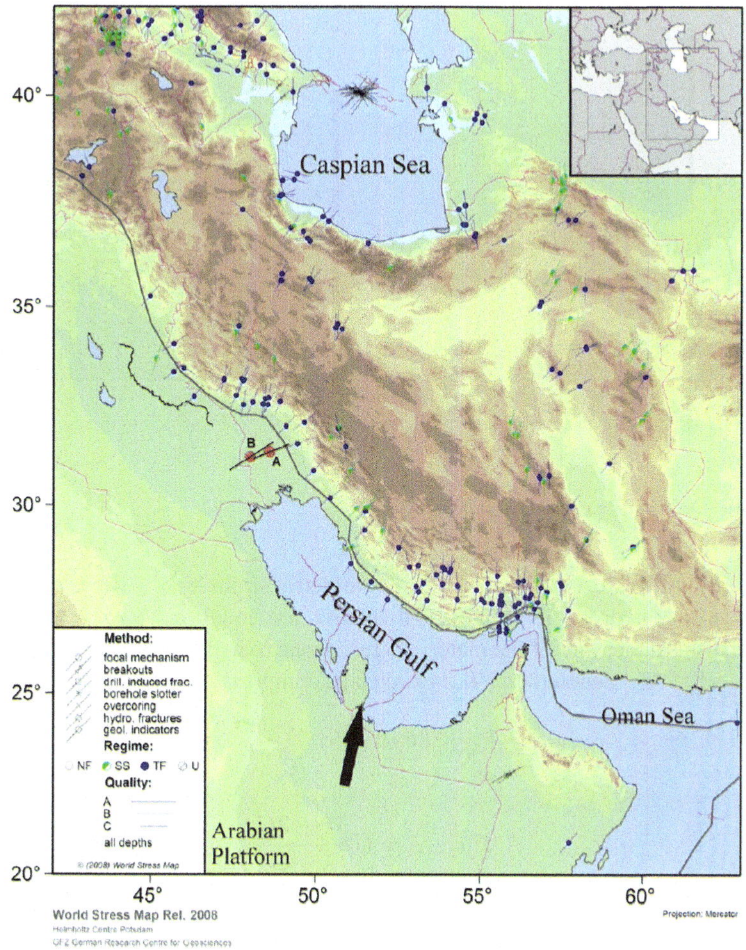

Fig. 2.9 Orientation of the maximum horizontal stress in Iran from the world map of the stress database

Typically, D' is greater than D ($D' > D$), resulting in negative values for both ε_y and ε_z. Equation 2.10 defines another important elastic parameter called Poisson's ratio. Poisson's ratio measures the lateral expansion relative to longitudinal contraction.

$$\vartheta = -\frac{\varepsilon_y}{\varepsilon_z} \tag{2.10}$$

The coefficients λ and G represent elastic moduli and are referred to as Lame parameters. G is also recognized as the modulus of rigidity or shear modulus, and it quantifies a material's resistance to shear deformation. Another significant elastic modulus is the bulk modulus 'K' (as described in Eq. 2.11). This modulus is defined

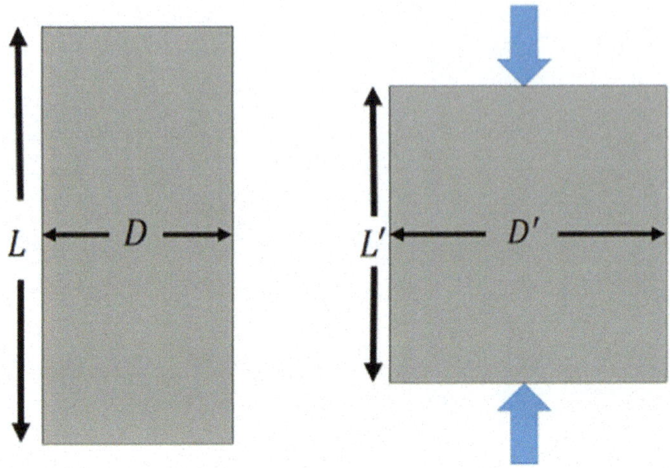

Fig. 2.10 Deformation caused by uniaxial stress

as the ratio of hydrostatic stress 'σ_p' to volumetric strain 'ε_{vol}' (where $\varepsilon_{vol} = \varepsilon_x + \varepsilon_y + \varepsilon_z$). In a hydrostatic scenario, all stresses are equal ($\sigma_p = \sigma_x = \sigma_y = \sigma_z$). '*K*' serves as a measure of the material's resistance to hydrostatic compression. The reciprocal of '*K*,' denoted as 1/*K*, is known as compressibility.

$$\mathrm{K} = \frac{\sigma_\mathrm{p}}{\varepsilon_\mathrm{vol}} = \lambda + \frac{2}{3}\mathrm{G} \tag{2.11}$$

$$\mathrm{E} = \frac{\sigma_x}{\varepsilon_x} = G\frac{3\lambda + 2G}{\lambda + G} \tag{2.12}$$

$$\vartheta = -\frac{\varepsilon_y}{\varepsilon_x} = \frac{\lambda}{2(\lambda + G)} \tag{2.13}$$

It can be observed from Eqs. 2.6–2.13 that when two of the following parameters are defined: *E*, *ν*, λ, *G*, and *K*, the remaining parameters can also be determined. The quantity H, represented as $H = \lambda + 2G$, is the uniaxial bulk modulus or oedometer modulus. The elastic moduli *E*, λ, *G*, *H*, and *K* are expressed in the same units of stress, such as Pa or psi. Some of the most valuable relationships between these moduli are provided in Table 2.2.

Example 2.4 Given the equations provided, calculate the value of the bulk modulus (K) and the compressibility (1/K) for a material with Lame parameters $\lambda = 5$ and $G = 3$.

Answer:
Using Eq. 2.11, we can calculate the bulk modulus (K):

$$K = \sigma_p/\varepsilon_\mathrm{vol} = \lambda + 2/3G$$

Table 2.2 Some relationships between elastic moduli (Zoback, 2010)

$\frac{\lambda}{\lambda+G} = 2\vartheta$	$K = \lambda\frac{1+\vartheta}{3\vartheta}$	$E = 3K(1-2\vartheta)$
$\frac{G}{\lambda+G} = 1-2\vartheta$	$K = \frac{2}{3}G\frac{1+\vartheta}{1-2\vartheta}$	$\mathrm{E} = 2\mathrm{G}(1+\vartheta)$
$\frac{\lambda+2G}{\lambda+G} = 2(1-\vartheta)$	$\mathrm{K} = \lambda + \frac{2}{3}G$	$\mathrm{E} = \frac{9KG}{3K+G}$
$\frac{3\lambda+2G}{\lambda+G} = 2(1+\vartheta)$	$\mathrm{K} = \frac{GE}{(9G-3E)}$	$E = G\frac{3\lambda+2G}{\lambda+G}$
$\frac{3\lambda+4G}{\lambda+G} = 2(2-\vartheta)$	$\frac{\lambda}{G} = \frac{2\vartheta}{1-2\vartheta}$	$E = \frac{\lambda}{\vartheta}(1+\vartheta)(1-2\vartheta)$
$\vartheta = \frac{3K-2G}{2(3K+G)}$	$\mathrm{H} = \mathrm{K} + \frac{4}{3}G$	$H = \lambda + 2G$
$\mathrm{H} = 3\mathrm{K}\frac{1-\vartheta}{1+\vartheta}$	$\mathrm{H} = 2\mathrm{G}\frac{1-\vartheta}{1-2\vartheta}$	$H = E\frac{1-\vartheta}{(1+\vartheta)(1-2\vartheta)}$

$$= 5 + 2/3(3) = 5 + 2 = 7$$

Therefore, the bulk modulus (K) for this material is 7.
To find the compressibility (1/K), we take the reciprocal of K:

$$1/K = 1/7$$

Hence, the compressibility for this material is approximately equal to 0.1429.

2.9 Construction of Geomechanical Model

The initial step in characterizing the in-situ stresses within a region involves determining the mechanical properties of the formation. Typically, the procedure for acquiring these parameters begins with laboratory tests on core samples, where their static values are measured. Subsequently, dynamic values are calculated using sonic and density logs, along with established relationships. These dynamic values are then correlated with the corresponding static values. This conversion of dynamic data to static values enables the utilization of all dynamic measurements from the logs in the geomechanical analysis of reservoirs. However, obtaining certain mechanical parameters of reservoirs, such as Biot's coefficient and bulk modulus, in the laboratory can be a complex task. Due to the limited availability of well cores, there's a growing trend to estimate geomechanical parameters like elastic constants and resistance parameters using petrophysical logs.

It's important to note that the relationships presented apply to materials exhibiting linear, isotropic, and homogeneous elastic behavior. Since rocks are often not isotropic or homogeneous, and they may not always demonstrate linear elastic behavior, the values obtained are generally close estimates and provide a preliminary perspective. Additionally, various relationships are provided for determining the elastic moduli of rocks, allowing the determination of other elastic constants with just two moduli, such as Young's modulus and Poisson's ratio.

Following the calculation of resistance parameters and static coefficients, the subsequent step involves determining the stress tensor and pore pressure within the hydrocarbon field. Understanding the type of stress variations is a crucial parameter for field development. In geologically complex areas, such as those with faults or folds, gathering information about geological structures aids in achieving geomechanical objectives more effectively.

To construct a precise mechanical model of the reservoir, a variety of data from diverse sources is essential. The level of access to this data, as well as the quality of its recording and processing, significantly influence the accuracy and validity of the constructed mechanical model. The necessary data for building an earth mechanical model can be categorized as follows:

- Well-logging data and the results of petrophysical interpretation, including parameters like porosity and the volume percentage of rock-forming minerals.
- Information related to well drilling, including records of mud flow, mud weight graphs, and drilling head data, such as drill head weight, rotation speed, and penetration rate into the formation.
- Interpretation results of visual diagrams, encompassing shear failure, induced tensile failure, and natural fractures within the formation.
- Detailed geological information, such as access to graphical well diagrams and well diameter charts.
- Results from laboratory tests on rock mechanics and data for estimating in-situ stresses, which include leakage tests and derived leakage data.

The algorithm for constructing a mechanical earth model is depicted in Fig. 2.11. The process of building the mechanical model commences with an examination of geological data, followed by the mechanical categorization of these layers based on geological indicators, including the minerals constituting the rock. Petrophysical logs and laboratory data aid in estimating rock elastic properties and fracture characteristics. Pore pressure is evaluated based on reservoir pressure tests and petrophysical logs. Subsequently, on-site stresses are calculated, beginning with the simplest part—overburden stress estimation—and proceeding to determine the minimum and maximum horizontal stresses. The most challenging aspect of the modeling process is estimating the maximum horizontal stresses, which necessitates access to a comprehensive dataset and simulating fractures generated on the wellbore wall during drilling operations (Verma et al., 2021).

2.9.1 Determination of Elastic Constants

Elastic constants encompass Poisson's ratio, Young's modulus, volume, and shear modulus, categorized into two primary groups: static and dynamic. These parameters are determined through the following methods:

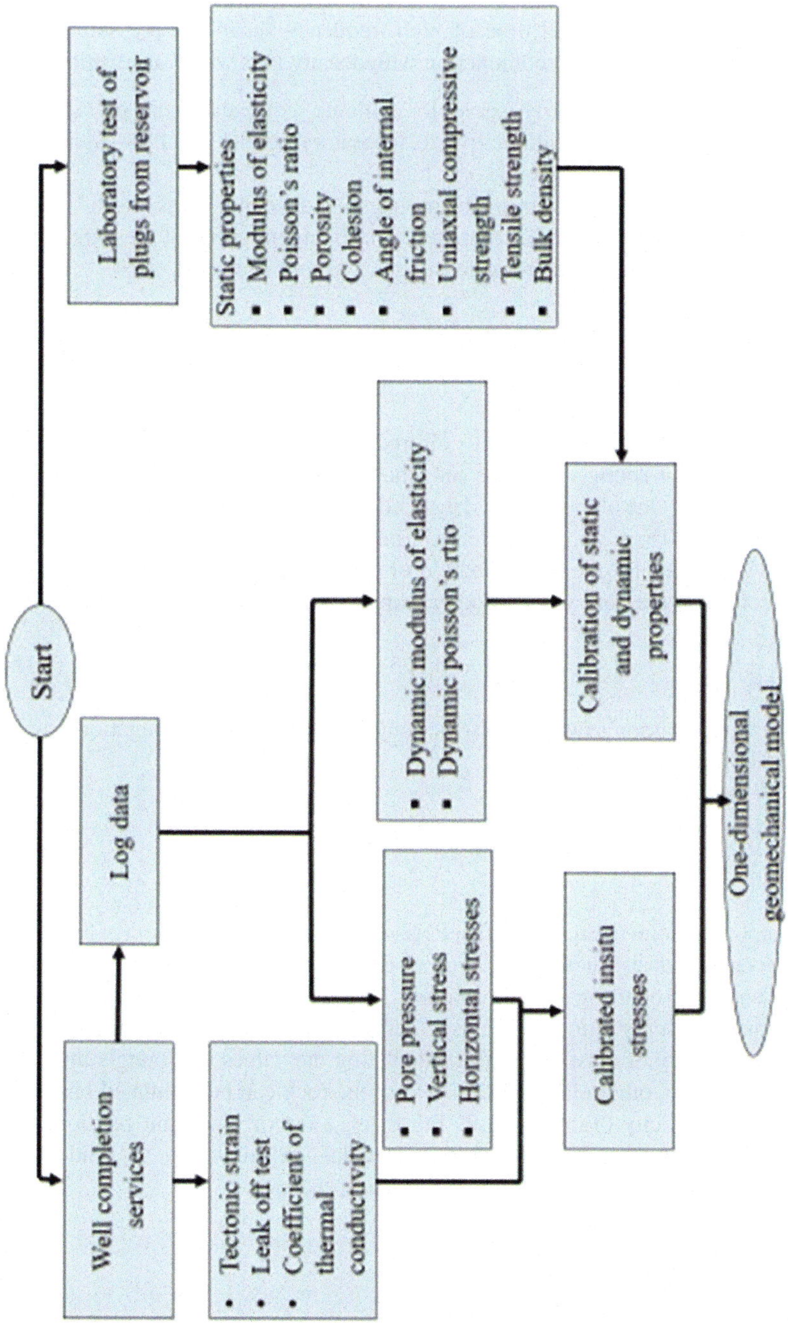

Fig. 2.11 Algorithm for building the mechanical model of the earth (Verma et al., 2021)

1. Measurement of mechanical properties, such as strain versus applied stress, or conducting uniaxial and triaxial compression tests on core samples.
2. Measurement of the travel time of high-frequency shear and pressure waves within the laboratory or in conjunction with density logs within the wellbore.

Laboratory results primarily serve to calibrate computational and indirect methods. A highly effective alternative to laboratory results involves calculating elastic properties using density and sonic logs.

According to the theory governing the propagation of sound waves in solid media, dynamic Young's modulus can be computed by utilizing the speed of pressure waves (V_p), shear waves (V_s), and the rock density (ρ_b), as outlined in Eq. 2.14:

$$E_{dyn} = \rho_b V_s^2 \left(\frac{3V_p^2 - 4V_s^2}{V_p^2 - V_s^2} \right) \quad (2.14)$$

In addition to this equation, various efforts have been undertaken to establish a connection between Young's modulus and other petrophysical parameters accessible from within a well. One of the most widely cited studies, conducted by Lacy, involved experiments on 400 core samples collected from the Gulf of Mexico. Lacy presented Eq. 2.15, which enables the determination of Young's dynamic modulus based on events recorded in pressure wave velocity graphs:

$$\mathrm{E_{dyn}} = 0.265\mathrm{V_p^{2.04}} \quad (2.15)$$

The dynamic Poisson's ratio can also be computed using wave propagation theory, as outlined in Eq. 2.16:

$$\vartheta_{\mathrm{dyn}} = 0.5\left(\frac{\mathrm{V_p^2 - 2V_s^2}}{\mathrm{V_p^2 - V_s^2}} \right) \quad (2.16)$$

The important point in Eq. 2.16 is that Poisson's ratio is only a function of the speed ratio of acoustic waves; therefore, direct access to pressure and shear sound wave values through well charting is very important. After calculating the values of Young's modulus and Poisson's ratio, other elastic parameters of the rock can be estimated based on the theory of elasticity. After calculating the values of Young's modulus and Poisson's ratio, other elastic parameters of the rock can be estimated based on the theory of elasticity (Table 2.1). In Table 2.3, a list of important relationships for determining static elastic constants is provided. In this table, Δt_c and t_s are respectively the duration of compression and shear waves.

Table 2.3 The relationship between static elastic constants and shear and pressure sound waves

Bulk modulus	$K_d = \rho\left(\frac{1}{\Delta t_c^2} - \frac{4}{3\Delta t_s^2}\right)$
Shear modulus	$G_d = \frac{\rho}{\Delta t_s^2}$
Elasticity modulus	$E_d = \frac{\rho\left(3-4\left(\frac{\Delta t_c}{\Delta t_s}\right)^2\right)}{\Delta t_s^2-\Delta t_c^2}$
Poisson's ratio	$\vartheta = \frac{0.5\left(\frac{\Delta t_s}{\Delta t_c}\right)^2-1}{\left(\frac{\Delta t_s}{\Delta t_c}\right)^2-1}$

2.9.2 Static Elastic Moduli

Dynamic elastic parameters can significantly differ from static elastic parameters. This discrepancy arises from the distinct methods used to calculate these parameters. In both dynamic and static measurement approaches, the rock undergoes loading. However, in dynamic measurement, rapid non-destructive loading occurs due to the passage of waves through the rock. Conversely, static measurement involves loading at a very slow rate, which can be destructive, leading to particle crushing, the formation of new cracks, or the expansion of existing ones. The slow loading rate and the occurrence of irreversible micromechanical phenomena within the rock contribute to a decrease in the measured values for static elastic parameters.

Dynamic coefficients cannot be directly employed in constructing a geomechanical model and must be converted to static values using empirical relationships. Creating a new relationship for each hydrocarbon field is often impractical due to the absence of laboratory cores. Existing empirical relationships from other regions serve as a suitable alternative, but the results should be adjusted based on local information.

Typically, there is no direct correlation between static and dynamic Poisson ratios. The static Poisson's ratio is usually assumed to fall between 0.7 and 1 times the dynamic Poisson's ratio. Additionally, volumetric and shear moduli can be converted using elastic modulus and Poisson's ratio, as specified in the formulas presented in Table 2.4. Consequently, determining the static elastic modulus allows for the calculation of other static constants.

Considering that static elastic parameters serve as critical criteria for most geomechanical studies, it becomes essential to extract their profiles within the well. Various models have been developed for this purpose. One of the commonly used empirical relationships for estimating static elastic modulus values based on dynamic data, known as the Morales relationship, is formulated as follows:

$$E_{sta} = (-2.21\varnothing + 0.963)E_{dyn} \tag{2.17}$$

In this context, E_{sta} and E_{dyn} represent the static and dynamic Young's moduli, respectively, while Ø denoting the porosity of the rock. Ameen et al., based on the outcomes of triaxial loading laboratory tests conducted on core samples from the

Table 2.4 Relationships of compressive strength, angle of internal friction, and adhesion strength of rock

Zoback	$\varphi = 18.532V_p^{0.5148}$
Lal	$C = \frac{5(V_p - 1)}{\sqrt{V_p}}$
Plumb	$UCS = 4.1089E_s + 2.28$
Lacy	$UCS = 0.278E_s^2 + 0.458E_d$
Najibi	$UCS = 11.05E_s^{0.66}$
Kahraman	$UCS = 9.95V_p^{1.21}$
Yasar and Ordogan	$UCS = 31.5V_p - 63.7$
Zoback	$UCS = \frac{\left(\frac{7682}{\Delta t_p}\right)^{1.82}}{145}$
Amani and Khalili	$UCS = 570.81e^{(-0.031 \times \Delta t_p)}$
Najibi	$UCS = 3.67V_p^{2.14}$

Arab reservoir layer in the Ghawar field, established a linear relationship between the static and dynamic Young's moduli, as presented in Eq. 2.18:

$$E_{sta} = 0.541E_{dyn} + 12.852 \tag{2.18}$$

Eissa and Kazi similarly introduced a linear relationship between static and dynamic elastic moduli by conducting studies on sedimentary rock samples, as outlined in Eq. 2.19:

$$E_{sta} = 0.74E_{dyn} - 0.82 \tag{2.19}$$

Van Heerden, through an examination of various relationships between static and dynamic Young's moduli across different rock types, asserted that a power-law relation, represented as Eq. 2.20, can be employed with considerable accuracy to estimate the static Young's modulus size:

$$E_{sta} = aE_{dyn}^b \tag{2.20}$$

In this regard, the values of '*a*' and '*b*' vary within the ranges of 0.097 to 0.152 and −1.485 to 1.388, respectively, for different types of rocks.

Lacy derived quadratic equations, denoted as 2.21 for sandstones and 2.22 for shales, using sonic data obtained from 600 core samples from the Gulf of Mexico in laboratory experiments.

$$E_s = 0.0293E_d^2 + 0.4533E_d \tag{2.21}$$

$$E_s = 0.0428E_d^2 + 0.2334E_d \tag{2.22}$$

Lacy also introduced a general equation, denoted as 2.23, which applies to shale, sandstone, limestone, and dolomite rocks.

$$E_s = 0.018E_d^2 + 0.422E_d \tag{2.23}$$

Ohne formulated Eq. 2.24 for rocks found in the Gulf of Mexico. In this regard, the dynamic Young's modulus is approximately 18 times greater than the static Young's modulus:

$$E_s = 0.0158E_d^{2.74} \tag{2.24}$$

Najibi et al., in their study involving core samples of Asmari and Sarvak limestone, introduced an exponential equation denoted as 2.25 for estimating the static Young's modulus.

$$E_{sta} = 0.014E_{dyn}^{1.96} \tag{2.25}$$

Unlike Young's modulus, there is a consensus among researchers that static and dynamic Poisson's ratios can be considered equal in the absence of experimental data indicating otherwise. Once the static Young's modulus is determined, other moduli such as volume and shear can be calculated.

2.9.3 Estimation of Rock Resistance Parameters

As mentioned earlier, rock resistance parameters such as compressive strength, internal friction angle, and rock adhesion resistance are crucial factors in the geomechanical analysis of hydrocarbon reservoirs. Determining these parameters can be accomplished using information obtained from drilling cores or by utilizing well data. However, due to the limited number of cores available, there is a significant risk associated with extrapolating the results to the entire reservoir layer.

On the other hand, determining these parameters through well-log measurements offers a faster and more generalizable approach. Therefore, providing relationships that estimate resistance parameters based on common well logs can be highly beneficial. Consequently, several empirical relationships have been proposed, and some of the most important ones are listed in Table 2.4.

2.9.4 Overburden Stress

The initial step in assessing the underground stress tensor involves determining the magnitude of the overburden stress. This stress arises from the weight of the rock column extending up to the ground level, and as a result, it can be computed by summing the weight of the overburden rock column using Eq. 2.26:

$$\sigma_v = \int_0^D \rho_b(z) g dz \tag{2.26}$$

where z, g, ρ_b, and 'D' represents the vertical depth from the ground surface, the acceleration due to the Earth's gravity, the rock density, and the desired final depth, respectively. The primary challenge in determining the overburden stress lies in obtaining a continuous profile of rock density at great depths beneath the surface. Typically, density logs are acquired at specific intervals, especially within reservoir layers, and are not available at deep subsurface depths.

Several methods exist to address this issue. The simplest approach involves using an average value for rock density, typically set at 2.6 gr/cc. A more sophisticated method is to employ geophysical observations to ascertain this value. Another proposed technique is establishing a relationship between density and sound wave velocity since sound speed logs are generally available at depths closer to the Earth's surface.

The relationship presented by Gardner, as outlined in Eq. 2.27, suggests an Exponent model connecting rock density and sound wave velocity:

$$\rho_b = 1.741 V_P^{0.25} \tag{2.27}$$

In this equation, ρ_b and V_P are expressed in grams per cubic centimeter (gr/cc) and kilometers per second (km/s), respectively. One of the most commonly employed relationships, widely used, is the Exponent equation denoted as 2.28, which relates density and depth:

$$\rho_b = \rho_{sur} + A_0(TVD - WD - AG)^{\eta} \tag{2.28}$$

In this context, the variables ρ_{sur}, TVD, WD, AG, η, and A_0 represent the following parameters: the density at ground level or sea floor, vertical depth, sea water height (in the case of offshore studies), the distance from the turntable to sea level, and the extrapolation parameters, respectively. ρ_{sur} is typically assumed to be 1.9 gr/cc in studies where there is no available data for rock density at the Earth's surface or seabed.

2.9.5 *Horizontal In-Situ Stresses*

In general, the horizontal stresses are caused by the two factors of overburden weight and tectonic activities; therefore, if the tectonic activities are weak, the main horizontal stresses are equal to each other and can be calculated based on the magnitude of Poisson's ratio of the reservoir rock from the main vertical stress. Based on this, Eaton presented Eq. 2.29 to calculate the number of stresses in the horizontal position:

Various methods have been presented to consider the effect of tectonic activities on the calculation of horizontal stresses. In the meantime, the calculation of the minimum horizontal stresses is simpler, while the calculation of the maximum horizontal stresses is sometimes associated with complications. One of the most famous models for estimating the size of horizontal stresses, known as the poroelastic model, was presented by Blanton and Olson. Based on the principles of rock mechanics, they presented a poroelastic model to calculate the horizontal stresses in the underground layers in the form of Eqs. 2.29 and 2.30:

$$\sigma_{\mathrm{hmin}} = \left(\frac{\nu}{1-\nu}\right)(\sigma_{\mathrm{v}} - \alpha \mathrm{p}) + \alpha \mathrm{p} + \frac{\mathrm{E}}{1-\nu^2}(\varepsilon_{\mathrm{h}} + \nu\varepsilon_{\mathrm{H}}) \tag{2.29}$$

$$\sigma_{\mathrm{Hmax}} = \left(\frac{\nu}{1-\nu}\right)(\sigma_{\mathrm{v}} - \alpha \mathrm{p}) + \alpha \mathrm{p} + \frac{\mathrm{E}}{1-\nu^2}(\varepsilon_{\mathrm{H}} + \nu\varepsilon_{\mathrm{h}}) \tag{2.30}$$

In this equation, α, E, ε_{H}, and ε_h correspond to Biot's coefficient, Young's modulus, and calibration coefficients for horizontal stresses based on tectonic strains, respectively. The primary challenge when employing this model lies in the estimation of the values for tectonic strains, ε_{h} and ε_{H}. To calculate these two calibration coefficients, data from leakage tests, developed leakage tests, minifrac, or microfrac tests can be utilized. These tests are all rooted in hydraulic fracturing operations and, by providing estimations of horizontal stress, contribute to determining the values of the tectonic parameters used in the aforementioned model. Additionally, the estimated stresses should be capable of predicting the fractures observed on the wellbore wall during drilling operations.

2.9.6 Biot Coefficient

Biot coefficient is usually determined through laboratory tests and its value varies from zero for tight rocks with zero porosity to one for porous rocks in shallow sedimentary basins. In the geomechanical studies of oil and gas reservoirs, due to the difficulty of performing the tests, they were omitted, and considering that the porosity values in carbonate rocks are in the range of 0.15–0.2, this coefficient is usually considered in the range of 0.6–0.8 (Biot, 1941, 1962).

2.9.7 In-Situ Stress Regime

In the basement, the stress conditions are typically characterized by three principal perpendicular stress components: the major stress (σ_1), intermediate stress (σ_2), and minor stress (σ_3). Depending on the prevailing tectonic conditions in a given region, these stresses can be represented by the vertical stress (σ_v), maximum horizontal stress (σ_{Hmax}), and minimum horizontal stress (σ_{hmin}).

In areas dominated by compressive tectonics, the maximum horizontal stress serves as the principal stress ($\sigma_1 = \sigma_{Hmax}$). In such conditions, when both the maximum and minimum horizontal stresses exceed the vertical stress ($\sigma_{Hmax} \geq \sigma_{hmin} \geq \sigma_v$), it leads to reverse faulting. Alternatively, if the vertical stress acts as the principal intermediate stress ($\sigma_{Hmax} \geq \sigma_v \geq \sigma_{hmin}$), and the disparity between the horizontal stresses is significant, it creates conditions conducive to strike-slip faulting.

Conversely, in regions where tensile tectonics are prevalent, the vertical stress becomes the primary principal stress ($\sigma_v \geq \sigma_{Hmax} \geq \sigma_{hmin}$). In such cases, gravitational forces can give rise to normal faults. Anderson introduced a classification for stress regimes, encompassing reverse, extensional, and normal stress regimes, known as fault stress regimes, based on these principles.

2.9.8 Estimation of Temperature Conditions

The thermal gradient within the Earth's subsurface plays a fundamental role in various geological processes, including the formation, maturation, and migration of hydrocarbon deposits, as well as influencing well-completion procedures and reservoir efficiency. Furthermore, the electrical and mechanical properties of rocks are temperature-dependent, underscoring the significance of accurate formation temperature estimation. The primary source of this thermal gradient is the radioactive decay of materials within the Earth. Interestingly, more than 70% of the temperature within the continental crust is generated at depths ranging from 10 to 20 km, while 96% of the temperature within the oceanic crust originates from radioactive substances within the sub-crustal regions of the ocean. The distribution of temperature is influenced by factors such as depth, the proportion of radioactive materials in the deeper layers, volcanic activity, regional tectonics, lithology or the thermal conductivity of materials, hydrology, and hydrothermal processes.

Formation temperature measurements are typically obtained through methods like well-temperature logging, drill stem tests, and the use of downhole temperature sensors.

2.10 Questions

Question 2.1 A rock weighing 23,400 N rests on a horizontal surface at the top of a 8.0-m-tall vertical pillar. The pillar's cross-sectional area is 0.40 m^2 and it is made of granite with a mass density of 2700 kg/m^3. Find the compressive stress at the cross-section located 6.0 m below the top of the pillar and the value of the compressive strain of the top 6.0-m segment of the pillar (Young's modulus for granite is Y = 4.5 $\times$ 10^{10} Pa).

Question 2.2 Calculate bulk modulus, shear moduls, elasticity modulus, poisson's ration.

Shear wave velocity: 80 m/s.

Compressive wave velocity: 120 m/s

Question 2.3 Log data recorded a saturated calcite reservoir. Sonic log was taken and average shear velocity and average compressive velocity are 65 and 110 m/s. Calculate static and dynamics elastic constants. Futhermore estimate rock resistance parameters (Use Plumb, Lacy, Najibi, Kahraman, Yasar and Ordogan, Zoback, Amani and halili, Najibi for calculating uniaxial compressive strength and compare to each other).

Question 2.4

A. What is the purpose of the Mohr–Coulomb criterion in describing failure behavior?
B. How is the Mohr–Coulomb criterion expressed mathematically?
C. What does it mean when Mohr's circle intersects the plotted line in the criterion?
D. What is the significance of the angle ϕ in the Mohr–Coulomb criterion?
E. How is the coefficient of internal friction related to the coefficient of internal friction?
F. What is the initial step in characterizing the in-situ stresses within a region?

References

Aleksandrov, K. (1961). The elastic properties of rock-forming minerals, II: Layered silicates. *Bulletin Academic Sciences USSR, Geophysics Series, 9*, 1165–1168.

Bai, Y., & Wierzbicki, T. (2010). Application of extended Mohr–Coulomb criterion to ductile fracture. *International Journal of Fracture, 161*(1), 1–20.

Biot, M. A., & Willis, D. G. (1957). *The elastic coefficients of the theory of consolidation.*

Biot, M. A. (1941). General theory of three-dimensional consolidation. *Journal of Applied Physics, 12*(2), 155–164.

Biot, M. A. (1955). Theory of elasticity and consolidation for a porous anisotropic solid. *Journal of Applied Physics, 26*(2), 182–185.

Biot, M. A. (1962). Mechanics of deformation and acoustic propagation in porous media. *Journal of Applied Physics, 33*(4), 1482–1498.

Bishop, A. W. (1959). The principle of effective stress. *Teknisk Ukeblad, 39*, 859–863.

Breckels, I. M., & Van Eekelen, H. A. M. (1982). Relationship between horizontal stress and depth in sedimentary basins. *Journal of Petroleum Technology, 34*(09), 2191–2199.

Byerlee, J. D., & Zoback, M. D. (1975). Permeability and effective stress: Geologic notes. *AAPG Bull, 59*(1), 154–158.

Chau, K. T., & Wong, R. (1996). Uniaxial compressive strength and point load strength of rocks. *International Journal of Rock Mechanics and Mining Sciences and Geomechanics Abstracts.*

Cheng, A.-D. (1998). On generalized plane strain poroelasticity. *International Journal of Rock Mechanics and Mining Sciences, 35*(2), 183–193.

Coussy, O. (2004). *Poromechanics.* John Wiley & Sons.

Detournay, E., & Cheng, A.-D. (1988). Poroelastic response of a borehole in a non-hydrostatic stress field. *International Journal of Rock Mechanics and Mining Sciences and Geomechanics Abstracts.* Elsevier.

Eaton, B. A. (1969). Fracture gradient prediction and its application in oilfield operations. *Journal of Petroleum Technology, 21*(10), 1353–1360.

Farmer, I. W. (2012). *Engineering behaviour of rocks.* Springer.

Gueguen, Y., & Bouteca, M. (1999). Mechanical properties of rocks: Pore pressure and scale effects. *Oil and Gas Science and Technology, 54*(6), 703–714.

Heidbach, O., Tingay, M., Barth, A., Reinecker, J., Kurfeß, D., & Müller, B. (2010). Global crustal stress pattern based on the World Stress Map database release 2008. *Tectonophysics, 482*(1–4), 3–15.

Hoek, E. (1990). Estimating Mohr–Coulomb friction and cohesion values from the Hoek-Brown failure criterion. *International Journal of Rock Mechanics and Mining Sciences and Geomechanics Abstracts.*

Hubbert, M. K., & Willis, D. G. (1957). Mechanics of hydraulic fracturing. *Transactions of the AIME, 210*(01), 153–168.

Ismail, M. J., Ettensohn, F. R., Handhal, A. M., & Al-Abadi, A. (2021). Facies analysis of the Middle Cretaceous Mishrif Formation in southern Iraq borehole image logs and core thin-sections as a tool. *Marine and Petroleum Geology, 133*, 105324.

Kahraman, S. (2001). Evaluation of simple methods for assessing the uniaxial compressive strength of rock. *International Journal of Rock Mechanics and Mining Sciences, 38*(7), 981–994.

Labuz, J. F., & Zang, A. (2012). Mohr–Coulomb failure criterion. *Rock Mechanics and Rock Engineering, 45*, 975–979.

Liu, J.-C., & Znidarčić, D. (1991). Modeling one-dimensional compression characteristics of soils. *Journal of Geotechnical Engineering, 117*(1), 162–169.

Martínez-Martínez, J., Benavente, D., & García-del-Cura, M. (2012). Comparison of the static and dynamic elastic modulus in carbonate rocks. *Bulletin of Engineering Geology and the Environment, 71*, 263–268.

Matthews, W. R. (1967). How to predict formation pressure and fracture gradient from electric and sonic logs. *Oil and Gas, 20*, 92–106.

Salemi, H., Iglauer, S., Rezagholilou, A., & Sarmadivaleh, M. (2018). Laboratory measurement of Biot's coefficient and pore pressure influence on poroelastic rock behaviour. *The APPEA Journal, 58*(1), 182–189.

Serra, O. (1989). *Formation MicroScanner image interpretation.*

Teymen, A., & Mengüç, E. C. (2020). Comparative evaluation of different statistical tools for the prediction of uniaxial compressive strength of rocks. *International Journal of Mining Science and Technology, 30*(6), 785–797.

Tingay, M. R., Hillis, R. R., Morley, C. K., King, R. C., Swarbrick, R. E., & Damit, A. R. (2009). Present-day stress and neotectonics of Brunei: Implications for petroleum exploration and production. *AAPG Bulletin, 93*(1), 75–100.

Verma, A. K., Deb, D., Dey, A. C., Roy, S., Singh, A. K., Avadhani, V. L. N., & Tiwari, R. R. (2021). Development of one dimensional geomechanical model for a tight gas reservoir. *Scientific Reports, 11*(1), 21433. https://doi.org/10.1038/s41598-021-00860-z

Wang, H. (2000). *Theory of linear poroelasticity with applications to geomechanics and hydrogeology* (Vol. 2). Princeton university press.
Zhang, J. (2011). Pore pressure prediction from well logs: Methods, modifications, and new approaches. *Earth-Science Reviews, 108*(1–2), 50–63.
Zoback, M. D. (2010). *Reservoir geomechanics*. Cambridge University Press.

Chapter 3
Static and Dynamic Tests

3.1 Introduction

Maintaining profitable production levels is of paramount importance when extracting resources from a reservoir. Matrix acidizing in carbonate reservoirs stands as a common practice to enhance production efficiency. The process of acidizing in porous spaces involves a dynamic, non-isothermal flow with chemical reactions. Once acid is injected into the formation, it initiates the dissolution of the rock. Consequently, the properties of the reservoir rock, namely porosity and permeability, undergo continuous changes over time. In the oil and gas industry, acid is frequently employed to remediate damage resulting from drilling operations, returning the well to the production cycle. Acid serves to boost production and injection rates by eliminating inorganic deposits, creating wormholes, and establishing new pathways around the well. Multiple factors come into play in the matrix acidizing process. One of the most critical considerations involves directing acid to regions characterized by low permeability. Another pivotal factor influencing the success of acidizing operations is the rate of reaction between the acid and the rock. To generate deep wormholes, a delicate balance must be struck: the reaction rate should be low, while the acid must remain potent enough to react with the rock over substantial distances. This chapter aims to optimize the choice of acid type, concentration, and volume through dissolution tests (Khalil et al., 2021).

3.2 Experiment Design

Experiment design techniques emerged in the 1980s and 1990s as a competitive advantage in Western countries and Japan, offering a means to enhance quality. The precise application of statistical test design methods facilitates the creation of

© The Author(s), under exclusive license to Springer Nature Switzerland AG 2024

E. Khamehchi et al., *Applied Matrix Acidizing of Carbonate Reservoir*,
Petroleum Engineering, https://doi.org/10.1007/978-3-031-58281-3_3

new products and the improvement of existing ones. These principles find application across diverse industries, including electronics, semiconductors, aerospace, automotive, medical equipment, food, pharmaceuticals, chemicals, and process industries.

To attain specific criteria in an experiment, it is imperative to define all the parameters that impact a process through careful experimental design. Two prevalent methods in experimental design are Factorial and Taguchi. A factorial experiment involves two or more factors, each with discrete values or levels, where experimental units encompass all conceivable combinations of these levels across all factors. This allows researchers to investigate the individual effects of each factor on the response variable, as well as the interactions between factors and their effects on the response variable. The Taguchi method aims to streamline the number of potential states by limiting the parameters involved. This approach yields comparable results to the factorial method but with fewer tests. However, for examining dissolution tests and reaction speed, where a thorough examination of influencing parameters is needed, it is not advisable to employ the Taguchi method. Table 3.1 presents the influencing parameters for selecting the optimal concentration and volume of acid. The choices for concentration and volume, as well as the ratio of rock to acid, are based on prior articles and research findings.

Table 3.2 outlines the design algorithm for the hydrochloric acid test. Following this table, the entire set of experiments concerning hydrochloric acid is structured. For 7.5% hydrochloric acid, the minimum stoichiometric ratio of rock to acid is 1:7. Consequently, tests numbered 1–15, which pertain to rock-to-acid ratios less than 1:7, have not been included in the investigation.

Table 3.1 Hydrochloric acid test design

Time (min)	Concentration (%)	Rock to acid ratio	The volume of acid in solution	Acid type
30	7.5	1:3	100	Hydrochloric acid
60	15	1:5		
90	20	1:7		
120	28			
360				

Table 3.2 Hydrochloric acid test design algorithm

Number	Rock to acid ratio	Concentration (%)	Time (min)	Dissolution (%)	pH
1	1:3	7.5	30	30.3	− 0.5
2	1:3	7.5	60	35	− 0.45
3	1:3	7.5	90	38.9	− 0.2
4	1:3	7.5	120	42.44	− 0.5
5	1:3	7.5	360	46.2	0.5
6	1:5	7.5	30	36.82	0.66
7	1:5	7.5	60	40.25	0.7
8	1:5	7.5	90	45.3	− 0.5
9	1:5	7.5	120	49.4	− 0.2
10	1:5	7.5	360	48.53	1.7
11	1:7	7.5	30	40	0.6
12	1:7	7.5	60	51	0.7
13	1:7	7.5	90	54.3	0.8
14	1:7	7.5	120	60	0.7
15	1:7	7.5	360	60.1	1.6
16	1:3	15	30	30.3	− 0.5
17	1:3	15	60	35	− 0.45
18	1:3	15	90	38.9	− 0.2
19	1:3	15	120	42.44	− 0.5
20	1:3	15	360	46.2	0.5
21	1:5	15	30	40.84	− 0.6
22	1:5	15	60	44.25	− 0.4
23	1:5	15	90	54.7	− 0.3
24	1:5	15	120	53.03	− 0.3
25	1:5	15	360	53.5	− 0.4
26	1:7	15	30	35.173	− 0.5
27	1:7	15	60	43.04	− 0.4
28	1:7	15	90	52.32	− 0.4
29	1:7	15	120	60	− 0.4
30	1:7	15	360	60.1	− 0.3
31	1:3	20	30	30.3	− 0.7
32	1:3	20	60	34.23	− 0.6
33	1:3	20	90	38.9	− 0.6
34	1:3	20	120	41.4	− 0.5
35	1:3	20	360	46	− 0.3
36	1:5	20	30	40.84	− 0.6

(continued)

Table 3.2 (continued)

Number	Rock to acid ratio	Concentration (%)	Time (min)	Dissolution (%)	pH
37	1:5	20	60	44.25	− 0.6
38	1:5	20	90	54.7	− 0.5
39	1:3	28	30	26.6	− 1.3
40	1:5	20	360	56.64	− 0.4
41	1:7	20	30	44.015	− 0.7
42	1:7	20	60	46.4	− 0.7
43	1:7	20	90	54.3	− 0.6
44	1:7	20	120	63.3	− 0.5
45	1:7	20	360	65.4	− 0.3
46	1:3	28	30	26.6	− 1.3
47	1:3	28	60	29.35	− 1.25
48	1:3	28	90	32.1	− 1.2
49	1:3	28	120	34.4	− 1.2
50	1:3	28	360	36	− 1.1
51	1:5	28	30	26	− 1.2
52	1:5	28	60	29.2	− 1.1
53	1:5	28	90	34	− 1
54	1:5	28	120	37.9	− 0.9
55	1:5	28	360	40	− 0.8
56	1:7	28	30	40	− 1
57	1:7	28	60	43	− 0.9
58	1:7	28	90	46	− 0.8
59	1:7	28	120	49	− 0.7
60	1:7	28	360	50	− 0.6

3.3 Investigating Influencial Parameters

3.3.1 Choosing the Type of Acid

The selection of acid for acidizing operations depends on various factors, including the rock type, the type of damage present, and the nature of the reservoir fluid. The nature of the reservoir rock plays a significant role in determining the appropriate acid type. Hydrochloric acid, a common mineral acid, finds extensive use in acidizing carbonate rocks. Its effectiveness in dissolving minerals like calcium and magnesium makes it the acid of choice for these reservoirs, typically used at a concentration of 15%. Furthermore, XRD test results indicate a substantial presence of calcite in the samples, further substantiating the suitability of hydrochloric acid for acidizing. In this research, the concentration of hydrochloric acid is initially optimized to minimize the number of required tests. Following the determination of the optimal concentration for hydrochloric acid, the study proceeds to examine and test combinations of this optimal concentration with other substances like alcohols, acetic acid, and various additives.

3.3.2 Concentration

Hydrochloric acid is employed within a concentration range of 5–37% by weight, with 15% by weight being the most prevalent concentration used. Lower concentrations are typically employed to clear sediments, salt plugs, and emulsions, while higher concentrations are utilized to create broader and more extensive pathways within the reservoir rock. Hydrochloric acid concentrations are categorized at levels of 7.5, 15, 20, and 28%. The optimal concentration for each of these categorized levels is determined by considering all factors influencing the dissolution test.

3.3.3 Rock to Acid Ratio

In the industry, various ratios of rock to acid, such as 1:3, 1:5, 1:7, 1:10, and 1:20, have been utilized. To ascertain the optimal rock-to-acid ratio, the results obtained from the XRD test, coupled with stoichiometric reaction considerations, are taken into account. Additionally, two ratios exceeding the stoichiometric ratio are considered to achieve maximum solubility.

3.3.4 Time

The measurement of test time is a crucial parameter in dissolution testing. To assess reaction kinetics and determine the optimal flow rate, solubility must be measured at various time intervals. Specifically, time intervals of 30, 60, 90, 120, and 360 min are employed to establish the optimal acid volume through comparative analysis, ensuring consistency in the time parameter across different scenarios.

3.4 Rotary Disk Reactor

Acidizing is a widely employed well-stimulation technique in the oil and gas industry, aimed at enhancing well performance. It involves a chemical reaction between acid and the rock's pores, which enlarges the pore space and facilitates the flow of reservoir fluids toward the well. This process leads to increased oil and gas production and heightened productivity. The selection and optimization of acid, as well as process design, are directly contingent on the speed of the acid-rock reaction, the reaction mechanism, and the influencing factors.

Research findings indicate that steps can be taken to manipulate the speed of the acid-rock reaction, prolong the reaction duration, and expand the scope of acidizing. Consequently, studies on the kinetics of acid-rock reactions under reservoir conditions hold significant importance. The intricacies of rock mineral composition and structure, the uneven distribution of pores and fractures, and the complex patterns of acid flow within the rock pores and fractures make the behavior of acid in the formation highly intricate. The reaction between acid and rock within the formation, transpiring under varying temperatures and pressures, is inherently heterogeneous. The acid is transferred to the rock's surface and reacts through either convection or diffusion. The reaction rate of the system hinges on mass transfer and surface reaction processes. To simulate acid-rock reactions more effectively and align them closely with the conditions in the formation, dynamic tests employ a disk-turning apparatus. This setup allows for the determination of rock dissolution rates, reaction rate constants, reaction orders, and diffusion coefficients. Geological and engineering variables represent two fundamental factors that influence acid-rock reactions. Geological parameters, such as temperature, rock permeability, rock porosity, and rock mineral type, remain constant factors in acid-rock reactions. Employing spinning disk experiments to assess reaction speed aids in creating a more precise model for stimulating damage across various radii. The use of a rotating disk simplifies the determination of the diffusion coefficient compared to traditional flooding tests. Nonetheless, it's important to note that the rotating disk may underestimate the diffusion coefficient because the rock surface is exposed to a substantial amount of fluid, which does not accurately represent the fluid volume-to-pore volume ratio. Early studies by Lund explored the impact of rotation speed on the dissolution rate of calcite and dolomite using this device. Lund's research from 1975 found that calcite reacts with 1 M hydrochloric acid approximately 650 times faster than dolomite at 25 °C. Alkattan et al. (1998) demonstrated that the dissolution rate of calcite crystals, limestones, and compacted calcite powders remained consistent at temperatures of 25, 50, and 80 °C. These limestones contained less than 1% clay by volume, although one limestone variant contained 16% quartz by volume. These findings suggested that the dissolution rate of calcium carbonate was not significantly influenced by variations in mineralogical composition. Herman and White (1985) also observed that the dissolution rate of dolomite remained relatively unaffected by mineralogical factors. Gdanski (1999) reported that carbonate reservoir rocks tended to exhibit lower reactivity under reservoir conditions compared to the reactivity data reported for calcite and dolomite by Lund et al. Taylor employed the rotating disk test method to investigate the principles governing acid-rock reactions and asserted that porosity correlated with the reaction rate. Sayed et al. investigated the acid-rock reaction involving dolomite and emulsified acid, proposing a method to regulate emulsified acid concentration. By adjusting the rate and residual acid concentration, Qiu et al. assessed the impact of reaction products on acid-rock reactions through experiments involving limestone and marble using a rotating disk setup. Dong et al. conducted a study on the influence of temperature on acid-rock reactions in dolomite. Yoo et al. explored the effects of residual acid concentration on the kinetics of acid-rock reactions in dolomite. Li Qin introduced a novel approach to directing acid to the entire reactor for simultaneous stirring and assessed the impact of various physical parameters on the kinetics of acid-rock reactions. In the field of acid-rock reactions, various studies have contributed valuable insights. Lund et al. (1975) observed that calcite reacts significantly faster with 1 M hydrochloric acid than dolomite at 25 °C, indicating differences in reactivity. Alkattan et al. (1998) found that dissolution rates for calcite, regardless of their form (crystals, limestones, powders), remained consistent across different temperatures (25, 50, 80 °C). Moreover, they noted that variations in mineralogy did not significantly affect the dissolution rate of calcium carbonate. Herman and White (1985) also reported that dolomite's dissolution rate remains relatively unaffected by mineralogical differences. Furthermore, Gdanski (1999) reported that carbonate reservoir rocks tend to exhibit lower reactivity under reservoir conditions compared to the reactivity data for calcite and dolomite reported by Lund et al. Taylor's research suggested a correlation between porosity and the rate of acid-rock reactions. Sayed et al. proposed a method to control the concentration of emulsified acid during acid-rock reactions. Additionally, Qiu et al. conducted experiments using a rotating disk to assess the impact of reaction products on acid-rock reactions, particularly on limestone and marble. Dong et al. investigated the effect of temperature on acid-rock reactions involving dolomite, while Yoo et al. studied how residual acid concentration influences the kinetics of acid-rock reactions for dolomite. Li Qin introduced a novel method for directing acid uniformly across the reactor and explored the effects of various physical parameters on the kinetics of acid-rock reactions.

3.4.1 Rotating Disk Theory

The kinetics of the reaction are influenced by the materials and the rock, and the overall reaction consists of three stages (Fig. 3.1). Acid or liquid diffusion in the design process is required for: The rate-limiting step is defined as the slowest of the three steps that control the overall kinetics of the reaction. If the rate of reactant consumption is faster than mass transfer, the mass transfer process is limited. If the mass transfer is faster, the surface reaction will be limited (Rashed et al., 2016).

3.4.2 How to Work

The acid-rock reaction system comprises a plate, an acid-rock reactor, and a preheating reactor. Within this setup, a core sample is positioned on the stirring plate. The core is rotated within the acid-rock reactor, with the stirrer motor spinning

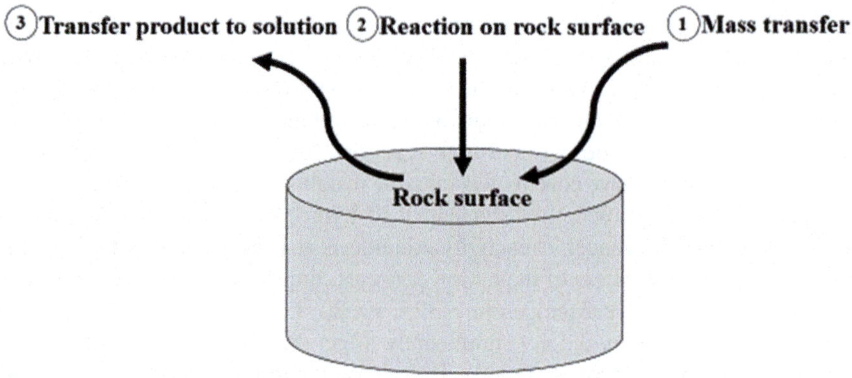

Fig. 3.1 Acid reaction with rock surface

the magnetic stirrer, which, in turn, rotates the stirrer rod. Pressure for the acid-rock reactor and reactor heating is supplied by the nitrogen cylinder and pressure relief valve within the pressure loading system. For a visual representation, refer to Fig. 3.2, which illustrates the rotary disk device, and Fig. 3.3, depicting a schematic diagram of the experimental apparatus. This experimental setup encompasses several key components, including a pressure loading system, a stirring system, a system for acid and rock reactions, a temperature control system, and a system for sampling and data collection. Together, these components facilitate the investigation of acid-rock reactions and their various parameters.

The temperature control system comprises a temperature controller and a heating cover. The required reaction temperature for the acid-rock test is adjusted using the temperature controller. The data acquisition system reads and records real-time parameters of the pressure sensor and the temperature control system. The sampling system includes a sampling valve and a sampling tube. The sampling valve is opened

Fig. 3.2 Rotary disk reactor (De Caprariis et al., 2015)

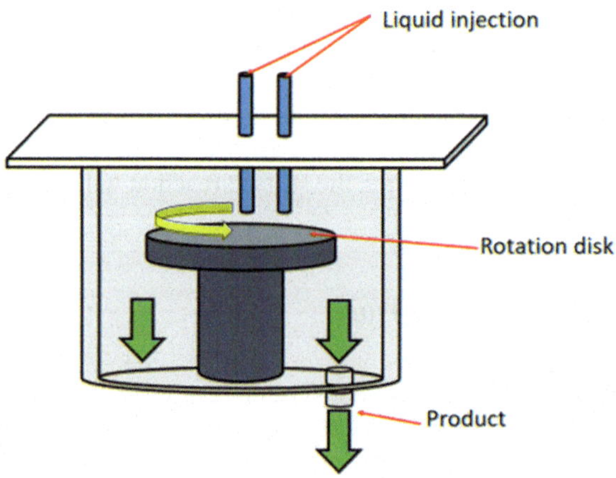

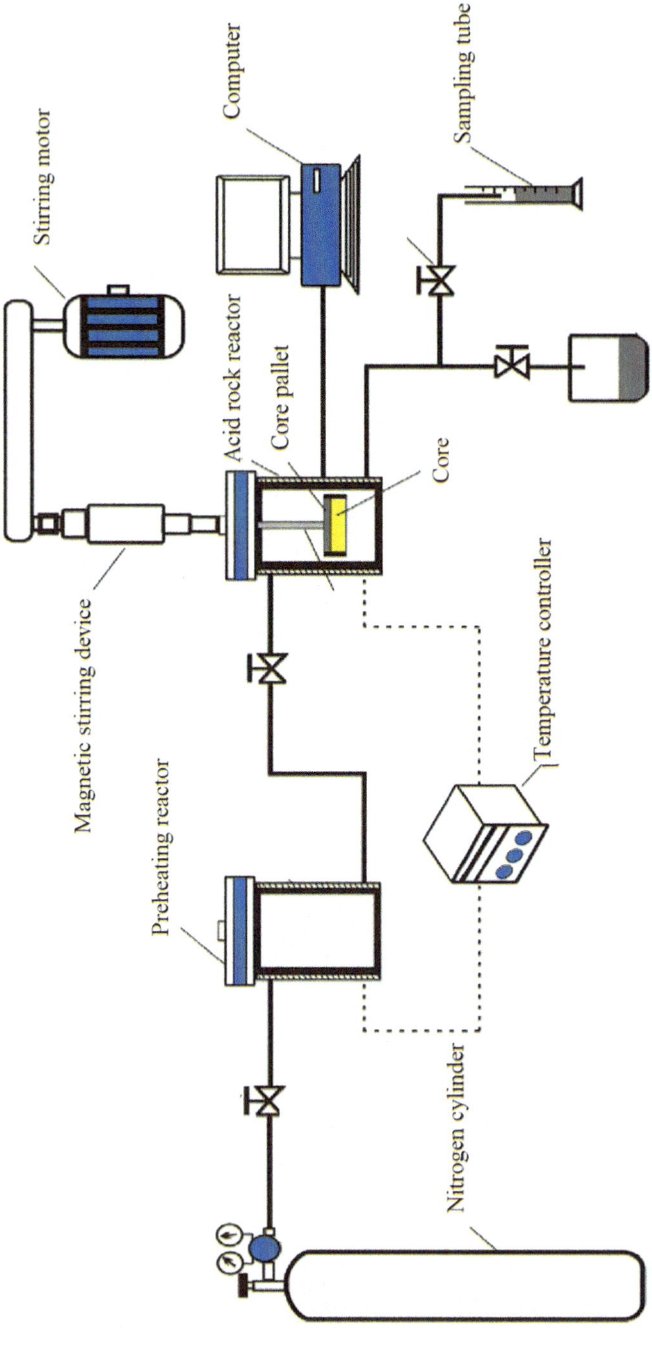

Fig. 3.3 Schematic diagram of a rotary disk reactor (Fan et al., 2023)

before the sampling process, and it is closed when a certain number of samples have entered the sampling tube.

The test steps are as follows:

1. The rock sample is dried in an oven before the test. The cross-section of the dried sample is photographed, and the diameter and thickness of the reaction surface are measured to calculate the area of the reaction surface.
2. The rotating disc device is assembled, ensuring that all parts are securely fastened.
3. Using a shrinkable rubber cylinder, the rock sample is attached to the plate and placed inside the acid-rock reactor.
4. The reactor is filled with an acid solution, and both the acid-rock reactor and rock samples are heated to the test temperature.
5. After adjusting the test speed, heated acid is injected into the acid-rock reactor using nitrogen. The rock and acid are given 300 s to react.
6. Sampling begins when the reaction time is reached. After sampling, stirring and gas flow are stopped, and the pressure is released. The rock sample is removed, washed, and photographed.
7. The acid-rock reactor is cleaned, and the acid is washed and discharged from the pipeline.

The reaction rate of the rock and rock can be calculated as the concentration of the acid solution decreases over time. The acid solution is allowed to react with the end surface of the core at a set temperature, pressure, and specified rotation speed. The concentration before and after the interaction of the acid solution is measured to determine the reaction rate of the rock and acid. When hydrochloric acid is used to acidize a typical oil and carbonate gas reservoir, the acid-dolomite chemical reaction formula is as follows:

$$4HCl + MgCa(CO_3)_2 \rightarrow CaCl_2 + MgCl_2 + 2CO_2 + H_2O \tag{3.1}$$

The initial concentration of C_o acid solution can be calculated as Eq. 3.2:

$$C_o = \frac{\rho_0 C_{NaOH} V_{NaOH-1}}{m_{HCl-1}} \tag{3.2}$$

The concentration of C_1 acid after the reaction can be calculated using Eq. 3.3, where C_o is the molar concentration of the acid solution (mol/L), ρ_o is the density of the acid solution (g/cm^3), C_{NaOH} is the molar concentration of the standard sodium hydroxide solution (mol/L), V_{NaOH-1} is the volume of the sodium hydroxide standard solution used (mL), and m_{HCl-1} is the mass of the acid solution (g).

$$C_1 = \frac{\rho_1 C_{NaOH} V_{NaOH-2}}{m_{HCl-2}} \tag{3.3}$$

The dynamic reaction rate of acid and rock can be calculated using Eq. 3.4, where C_1 is the molar concentration of residual acid (mol/L), V_{NaOH-2} is the volume of standard sodium hydroxide solution used (mL), ρ_1 is the density of residual acid (g/cm^3), and m_{HCl-2} is the mass of residual acid (g).

$$J = 10^{-3} \times \left(\frac{C_0 - C_1}{t_1 - t_2} \right) \frac{V}{S} \tag{3.4}$$

where J represents the dynamic acid-rock reaction rate in (mol/cm^2·s), V is the volume of the acid solution involved in the reaction in (mL), t_1 is the start time of the acid-rock reaction in seconds, t_2 is the end time of the reaction in seconds, and S stands for the contact surface of the acid-rock reaction in $\left(cm^2\right)$. Figure 3.4 illustrates the impact of temperature, disk rotation speed, and acid concentration on the speed of the acid-rock reaction. An increase in any of these three parameters leads to an enhanced reaction speed, demonstrating a direct correlation between these factors and the reaction rate (Fan et al., 2023; Taylor & Nasr-El-Din, 2009).

Example 3.1

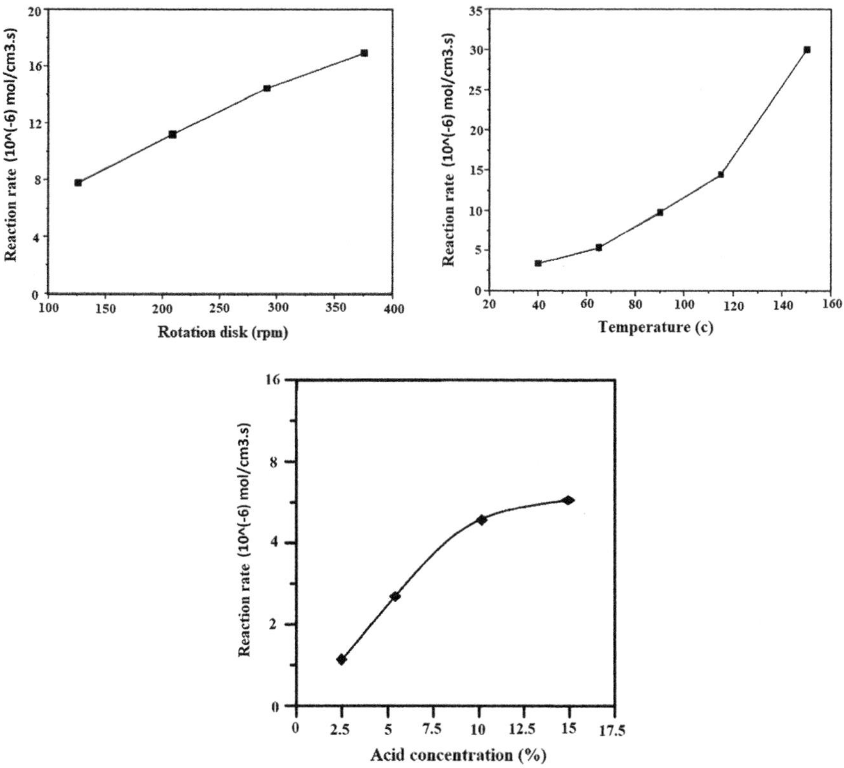

Fig. 3.4 Effect of temperature, rotary disk speed, and concentration on reaction rate (Fan et al., 2023)

A. The cross-section of a rock sample for an acid-rock test is photographed, revealing a diameter of 5 cm and a thickness of 2 cm determine the area of the reaction surface.
B. The acid-dolomite chemical reaction is represented by:

$$4HCl + MgCa(CO_3)_2 \rightarrow CaCl_2 + MgCl_2 + 2CO_2 + H_2O$$

Using the given equations, calculate the initial concentration of the acid solution with the following parameters:

$$\rho_0 = 1.3\frac{g}{cm^3}, \quad C_{NaOH} = \frac{0.08\ mol}{L}, \quad V_{NaOH-1} = 45\ mL, \quad m_{HCl-1} = 25\ g$$

C. Using the given equations, determine the concentration of acid after the reaction with the following parameters:

$$\rho_1 = 1.1\ \frac{g}{cm^3}, \quad C_{NaOH} = \frac{0.08\ mol}{L}, \quad V_{NaOH-2} = 18\ mL, \quad m_{HCl-2} = 12\ g$$

D. Calculate the dynamic acid-rock reaction rate with the following parameters:

$$t_1 = 350\ s, \quad t_2 = 150\ s, \quad V = 180\ mL$$

Answer:

A. The area of the reaction surface (A) is calculated using the formula for the area of a circle: $A = \frac{\pi d^2}{4}$

Substituting the given values: $A = \frac{\pi \times (5\ cm)^2}{4} = 19.63\ cm^2$

B. $C_0 = \frac{1.3 \times 0.08 \times 45}{25} = 0.1872\ mol/L$

C. $C_1 = \frac{1.1 \times 0.08 \times 18}{12} = 0.132\ mol/L$

D. $j = 10^{-3} \times \frac{(0.1872 - 0.132)}{(350 - 120)} \times \frac{180}{19.63} = 2.2 \times 10^{-6}\ mol/\left(cm^2 \cdot s\right)$

3.4.3 *Effect of Rock Porosity*

Rock porosity is often overlooked when evaluating acid reaction rates. However, it plays a significant role, as higher porosity results in a larger surface area for the rock disc involved in the reaction. Since the rate of acid dissolution is closely tied to the surface area of the rock disc, it follows that higher porosity leads to an increased dissolution rate. Conversely, rocks with low porosity tend to exhibit a minimized effect on this aspect of the reaction (De Caprariis et al., 2015).

3.4.4 *Rock Mineralogical Effect*

There is a common assumption that acid reactions with limestone reservoir rocks occur much more rapidly than with dolomite reservoir rocks. However, Taylor et al. were among the first to challenge this assumption, revealing that it does not always hold true. They pointed out that impurities commonly present in these rocks can alter the situation. Specifically, even minor clay impurities in limestone reservoir rocks can decrease the acid dissolution rate by up to 25%, effectively aligning the acid reactivity of these rocks with that of pure dolomitic rocks (De Caprariis et al., 2015; Taylor et al., 2006).

3.4.5 *Effect of Additives on Reaction Rate*

Taylor et al. investigated the impact of common acid additives on the reaction rate. Their findings revealed that the introduction of polymers can transform the acid-rock reaction, shifting it from being limited by mass transfer to being limited by surface reaction when dealing with calcite and dolomite. When iron(III) is present at a concentration of 5000 mg per liter, it results in the surface precipitation of iron(III) hydroxide on both calcite and dolomite. This deposit tends to have an inhibitory effect on the rate of rock dissolution, particularly at low rotational speeds. Furthermore, the addition of 12 g of citric acid per liter was found to reduce the dissolution rate of calcite by an average of 9%. This reduction is likely attributed to the formation of calcium citrate on the rock's surface (De Caprariis et al., 2015).

3.5 Hydrochloric Acid

Hydrochloric acid is a commonly employed mineral acid in the acidizing of carbonate reservoirs. It boasts a significantly higher acid strength when compared to other acids. This elevated acid strength renders hydrochloric acid more proficient in dissolving rock formations (Harris et al., 1966).

3.5.1 *Stoichiometry, Kinetics, and Equilibrium Constant*

The stoichiometry of a reaction defines the precise quantitative relationship between reactants and products. It specifies the proportion and quantity of each compound required in the reaction. The rate at which each product is consumed is determined by the reaction kinetics, which measures the speed of the reaction. A faster reaction results in a shorter reaction time. When the quantity of materials used is not optimized based on stoichiometric relationships, the reaction speed remains suboptimal. This, in turn, leads to a higher consumption of both materials and time to achieve the same reaction outcome. Table 3.3 provides information on the reactions between acid and calcite, dolomite, and siderite minerals (Chacon & Pournik, 2022; Economides & Nolte, 1989).

In acidizing, stoichiometry is expressed in terms of mass and volume dissolution power. Mass dissolution power (β) is the amount of mineral that is dissolved by a specific mass of acid. This parameter is defined as Eq. 3.5.

$$\beta = \frac{\vartheta_{mineral} MW_{mineral}}{\vartheta_{acid} MW_{acid}} \frac{(\text{mass of acid})}{(\text{mass of solutuon})} \tag{3.5}$$

Here, ϑ represents the stoichiometric coefficient, while $MW_{mineral}$ and MW_{acid} denote the molecular masses of the mineral and acid, respectively. The stoichiometric coefficients for common acidizing reactions can be derived from the reaction equations listed in Table 3.3.

The volumetric dissolution power (X) is expressed as the volume of mineral dissolved by a volume of acid. Equation 3.6 illustrates the volumetric dissolution power.

$$X = \beta \frac{\rho_{acidsolution}}{\rho_{mineral}} \tag{3.6}$$

The parameter ρ represents the density.

The reaction between the acid and the mineral takes place when the acid reaches the mineral surface through the processes of diffusion or convection from within the solution. The ultimate rate of acid consumption or mineral dissolution is dependent on the rate of acid transfer to the mineral surface and the actual rate of reaction occurring on the mineral surface. In general, the speed of the reaction depends on the concentration of acid and minerals within the rock. However, since the amount of minerals remains constant, it is often disregarded. Equation 3.7 represents the relationship between the reaction rate, acid concentration, and the surface area of the reaction.

Table 3.3 Chemical reactions in acidizing

Acid	Mineral	Reaction
HCl	Calcite	$2HCl + CaCO_3 \rightleftarrows CaCl_2 + H_2O + CO_2$
	Dolomite	$4HCl + CaMg(CO_3)_2 \rightleftarrows CaCl_2 + MgCl_2 + 2H_2O + 2CO_2$
CH3COOH	Calcite	$2CH_3COOH + CaCO_3 \rightleftarrows Ca(CH_3COO)_2 + H_2O + CO_2$
	Dolomite	$4CH_3COOH + MgCa(CO_3)_2 \rightleftarrows Mg(CH_3COO_2)_2 + Ca(CH_3COO_2)_2 + 2H_2O + 2CO_2$
HCOOH	Calcite	$2HCOOH + CaCO_3 \rightleftarrows Ca(COOH)_2 + H_2O + CO_2$

$$-R_A = E_f C_A^{\alpha} S_B \tag{3.7}$$

Here, R_A represents the rate of acid formation in moles per second (mole/s), E_f is the reaction rate constant in $\frac{\text{mol A}}{m^2-s-\left(\frac{\text{mol A}}{m^3}\right)^{\alpha}}$, C_A is the concentration of acid A in the surface reaction, and α is the reaction order. While reaction kinetics is related to the speed of a chemical reaction, the chemical equilibrium constant determines the direction and extent of the reaction. For a reaction under equilibrium control, equilibrium constants are used to relate the concentrations of reactants and products. The equilibrium constants are obtained from the thermodynamic properties of the substances involved in the reaction. Consider a simple reaction under controlled conditions where A and B are the reactants, and C and D are the products of the reaction. Equation 3.8 illustrates the relationship between reactants and products.

$$aA + bB \leftrightarrow cC + dD \tag{3.8}$$

If the system is in an ideal state, then the equilibrium constant is defined based on the concentration according to Eq. 3.9.

$$K_{eq} = \frac{[C]^c[D]^d}{[A^a][B^b]} \tag{3.9}$$

K_{eq} is the equilibrium constant, and expression [] expresses the concentration of each component (Economides & Nolte, 1989).

Example 3.2 Consider an acidizing reaction between formic acid (HCOOH) and calcite ($CaCO_3$) represented by the following chemical equation:

$$2HCOOH + CaCO_3 \leftrightarrows Ca(COOH)_2 + H_2O + CO_2$$

A. Use Eq. 3.5 to calculate the mass dissolution power (β) for this reaction. Given that $\vartheta_{mineral} = 1$, $\vartheta_{acid} = 2$, $MW_{mineral} = 100$ g/mol, $MW_{acid} = 46$ g/mol, and the mass of acid in the solution is 60 g.
B. Determine the mass of calcite that can be dissolved if the total mass of the solution is 150 g.
C. Using the calculated mass dissolution power from part A, find the volumetric dissolution power (X) using Eq. 3.6. Assume the density of the acid solution ($\rho_{acidsolution}$) is 1.1 g/mL and the density of calcite ($\rho_{mineral}$) is 2.71 g/cm^3.
D. The rate of acid formation (R_A) for the reaction is given by Eq. 3.7. For the reaction $E_f = 0.003 \frac{\text{mol A}}{m^2-s-\left(\frac{\text{mol A}}{m^3}\right)^{\alpha}}$, $C_A = 0.015$ mol/m^3, and $\alpha = 0.75$. Calculate R_A using these parameters.
E. If the equilibrium constant (K_{eq}) for the given reaction is 0.05 and the concentrations are [HCOOH] = 0.025 M, [$CaCO_3$] = 0.02 M, [$Ca(COOH)_2$] = 0.015 M, determine if the system is at equilibrium or not.

Answer:

A. $\beta = \frac{(1\times100)}{(2\times46)} = 1.09g/(gacid)$

B. Mass of calcite $= \beta\times$ (mass of acid solution) $= 1.09\,\frac{g}{gacid} \times 150\text{g} = 163.5\text{g}$

C. $X = \beta\frac{\rho_{\text{acid solution}}}{\rho_{\text{mineral}}} = 1.09\frac{g}{g\ acid} \times \frac{1.1\ g/mL}{2.71\ g/cm^3} = 0.44\frac{cm^3}{g\ acid}$

D. $R_A = 0.003\ \text{mol}/(\text{m}^2\ \text{s}\ (\text{mol}/\text{m}^3)^{0.75}) \times (0.015\ \text{mol}/\text{m}^3)^{0.75} \times S_B = -7.8 \times 10^{-5} \times S_B$

E. $K_{eq} = \frac{[Ca(COOH)_2]}{[HCOOH]^2} \times [H_2O] \times [CO_2]$

3.5.2 *Determination of Time and Ratio of Rock to Acid*

3.5.2.1 Solubility Test

When acids come into contact with calcium carbonate, there is an equilibrium between the reaction products and acidic species. This equilibrium is represented by Eq. 3.10.

$$H^+ + CaCO_3 \leftrightarrow Ca^+ + CO_2 + H_2O \tag{3.10}$$

In static conditions, the completion of the reaction depends on the concentration of carbon dioxide in the solution. This is because, as shown in Fig. 3.5, the carbon dioxide formed at higher pressures of 7 MPa remains in the solution, leading to the production of H_2CO_3. This formed acid has a buffering effect. In this pH range, acids cannot be completely decomposed. Therefore, in the conditions at the bottom of the well, acids can never be consumed 100%. For this reason, to conduct the solubility test, the system is considered closed.

To perform the solubility test and determine the end time of the reaction and the optimal ratio of rock to acid, it is necessary to dissolve a specific mass of rock in a certain volume of acid. For this purpose, first, a thin section of the rock reservoir with specific size, dimensions, and geometric shape (Fig. 3.6) is prepared (Garrouch & Jennings, 2017).

Then, this thin section is weighed and placed in containers with different acid concentrations. These bottles are then put in an oven at a temperature of 100 °C for various time intervals: 30, 60, 90, 120, and 360 min. In this experiment, different ratios of rock to acid were considered, including 1:3, 1:5, and 1:7. The mass of dissolved rock is measured at the end of each time interval. Figures 3.7, 3.8 and 3.9 depict samples before and after the solubility test.

One of the crucial factors in determining the optimal volume and time is the solubility of rock in acid. Increasing the acid volume leads to the dissolution of more calcite in the pores. However, it's important to note that there's a limit to this volume, as it only needs to dissolve all the calcite in the sample. Using higher volumes beyond this point doesn't increase the dissolution rate and merely consumes more acid.

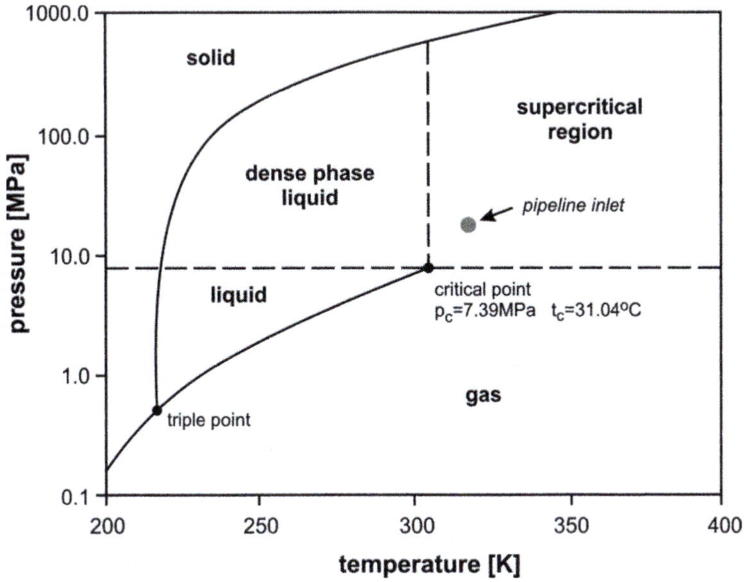

Fig. 3.5 The Phase diagram of carbon dioxide (Witkowski et al., 2014)

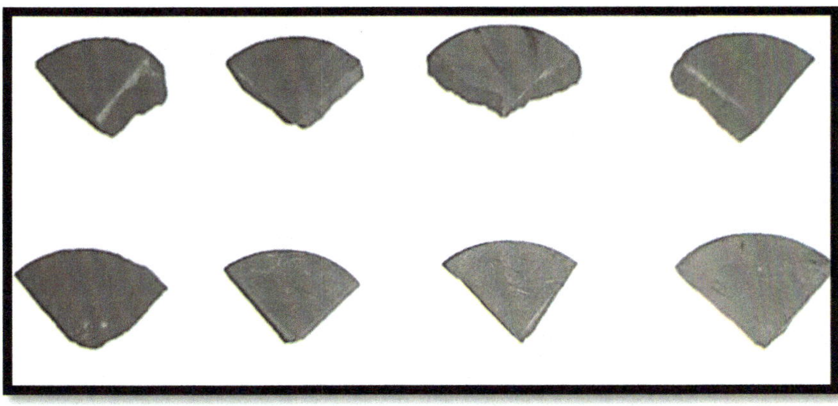

Fig. 3.6 An example of thin rock sections

Therefore, optimizing the time and rock-to-acid ratio can enhance the effectiveness of acidizing operations and reduce costs.

Based on the XRD test results, approximately 60% of the reservoir rock comprises calcite. Consequently, in the dissolution test, solubility is investigated up to a maximum of 60%. Figures 3.10, 3.11 and 3.12 depict the solubility of calcite over time for hydrochloric acid with different concentrations.

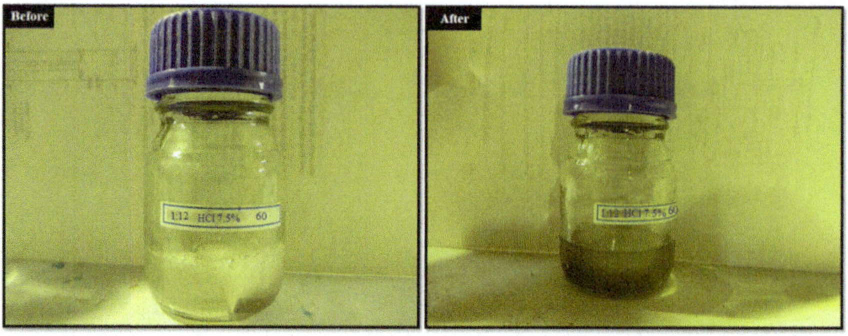

Fig. 3.7 Before and after placing the thin section of the plug in 7.5% HCl acid with a ratio of 1:5 for 60 min

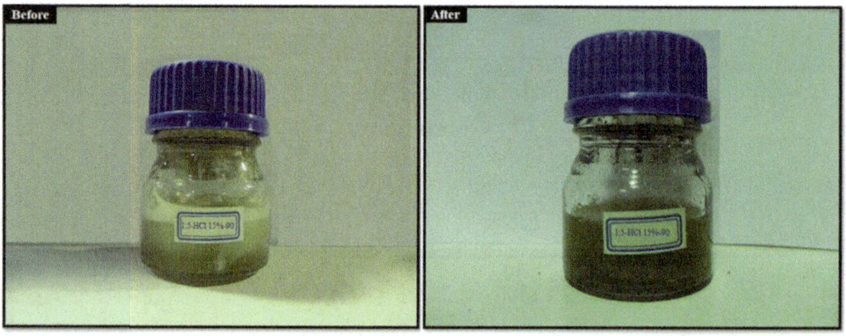

Fig. 3.8 Before and after placing the thin section of the plug in 15% HCl acid with a ratio of 1:3 for 90 min

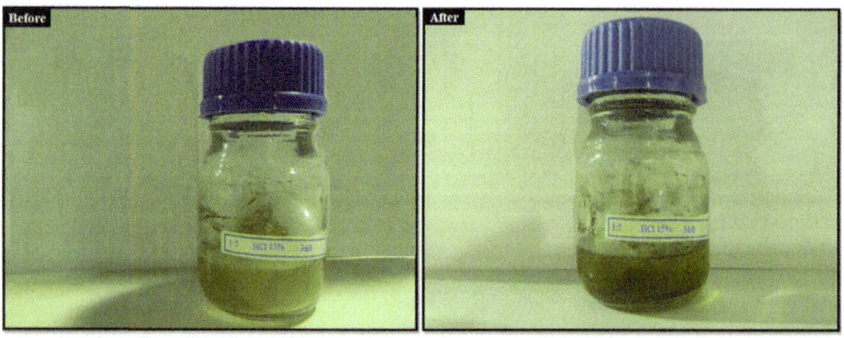

Fig. 3.9 Before and after placing the thin section of the plug in 15% HCl acid with a ratio of 1:3 for 360 min

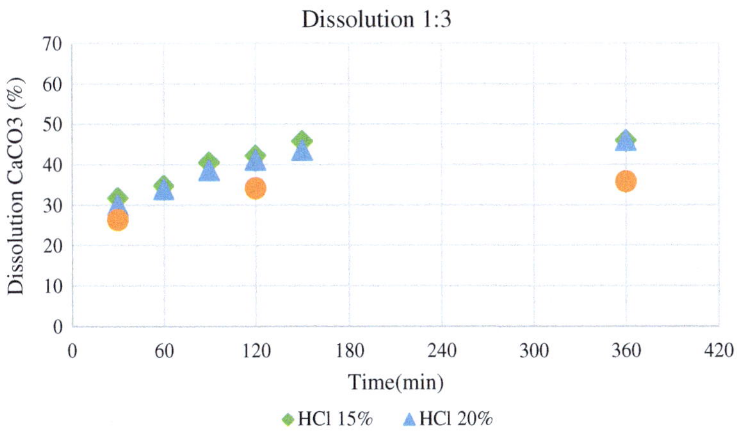

Fig. 3.10 Solubility of calcite in the presence of acid with different concentration in constant volume 1:3

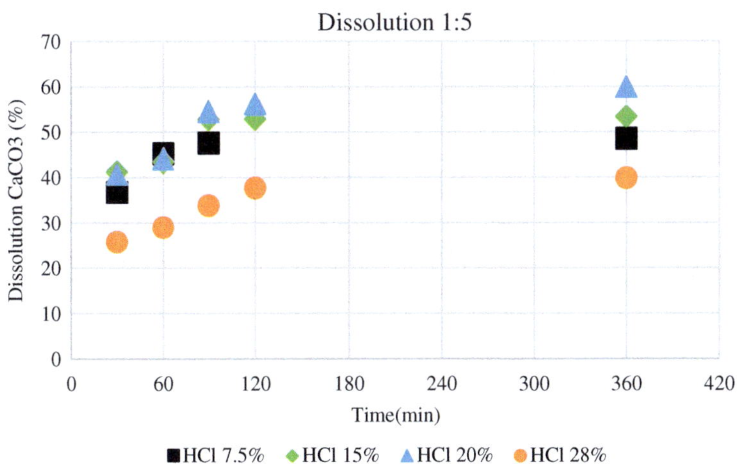

Fig. 3.11 Solubility of calcite in the presence of acid with different concentrations in constant volume 1:5

As observed in Figs. 3.10, 3.11 and 3.12, the dissolution rate tends to decrease with increasing acid concentration. Notably, 28% HCl exhibits lower solubility compared to other acids across all ratios, while the remaining acids display nearly identical solubility patterns. To further validate these findings, experiments were replicated using 28% HCl, which once again confirmed its lower solubility relative to other acids. The dissolution of carbonate core samples using acids at various concentrations was conducted within a closed environment, simulating reservoir conditions. In such closed systems, the acid-rock reaction isn't fully complete, and the reaction progresses in a balanced manner, often leading to reactions in the reverse direction.

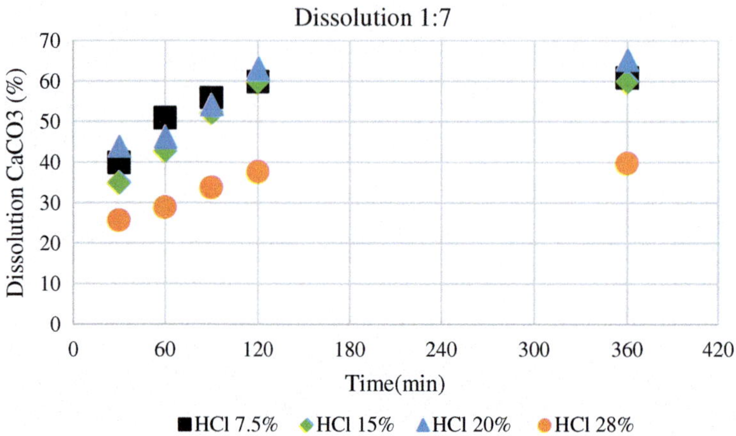

Fig. 3.12 Solubility of calcite in the presence of acid with different concentrations in constant volume 1:7

Conversely, in this closed environment, as acid concentration increases and the acid grows stronger, there is a reduced extent of reaction between the acid and the rock. The results obtained from the dissolution tests make it evident that as the acid strength increases, solubility decreases due to the closed environment and the reaction reaching equilibrium. To enhance the reliability of these findings, additional tests were conducted using the 28% acid sample, which yielded results consistent with the previously mentioned observations, reaffirming their validity. Given that more dissolution can be achieved with 28% HCl, it may seem tempting to increase the volume of acid, albeit at higher costs. However, it's important to note that the maximum solubility was achieved with weaker, less hazardous, and more cost-effective acids. Therefore, opting for a weaker acid with a higher volume would be a more economically and operationally sound choice compared to employing a strong acid with a larger volume.

Furthermore, by comparing the three graphs, it was discerned that for hydrochloric acid with concentrations of 7.5, 15, and 20%, the maximum solubility was reached at the 1:7 ratio. Consequently, the 1:7 ratio is recommended as the optimal ratio for these acids.

3.5.2.2 pH

Another parameter that is effective in choosing the optimal volume and time is pH after the acid-rock reaction. When discussing pH changes, if the reaction is endothermic, meaning it absorbs heat, higher temperatures will lead to a greater extent of separation, resulting in a decrease in pH. Conversely, if the reaction is exothermic, it will raise the pH. Measuring pH serves the purpose of determining when the reaction has concluded and to what degree it has progressed. For example, at the outset of the experiment, the HCl solution exhibits a low pH. However, as the reaction commences, the acid is consumed, leading to an increase in pH. After a certain point, the pH ceases to rise, indicating the end of the reaction and the depletion of all acids. In other words, the acid's strength becomes insufficient to dissolve the rock, and the reaction ceases. Therefore, it is essential to analyze the pH of the acid both before and after the dissolution test.

Strong acids are represented by the symbol HA, where H denotes a hydrogen atom and A represents an electronegative atom or species to which the hydrogen is attached. Strong acids are proficient proton donors and can remain in aqueous solution. They are classified as highly corrosive substances capable of producing hazardous fumes. Notable characteristics of strong acids include:

1. A strong acid possesses both the lowest pH and the highest pKa values. Strong acids are widely recognized as some of the most corrosive chemical compounds. Another characteristic of strong acids is their complete ionization when placed in water.
2. The ability of a solution to conduct electricity hinges on its ion concentration. In comparison to weak acids, strong acids have a higher ion concentration, making their solutions more effective electrical conductors.
3. Strong acids often function as catalysts in chemical reactions.

pH serves as a metric for gauging the acidity or alkalinity of solutions. According to various studies and reference materials, the pH value typically falls within the range of zero to 14 at room temperature. However, it's worth noting that some sources also mention the possibility of negative pH values or values exceeding 14.

According to Eq. 3.11, the pH of the solution can be less than zero if the hydronium ion concentration is greater than 1 mol/L.

$$\mathrm{pH} = -\log[\mathrm{H_3O^+(aq)}] \tag{3.11}$$

By checking the molarity for all 4 concentrations of 7.5, 15, 20, and 28, which are equal to 2, 4.11, 5.47, and 7.67 respectively, by inserting negative pH into the above equation, it can be seen that. Table 3.4 shows the acids used before the experiment.

The pH becomes increasingly negative in the presence of stronger acids, and its change is less pronounced due to their high acidity. Therefore, in situations where the pH remains negative throughout the reaction, it indicates a high concentration of acid. Figure 3.13 shows a continuous plot of pH variation over time for hydrochloric acid at concentrations of 15, 20, and 28%, with a rock/acid ratio of 1:3.

Figure 3.13 clearly illustrates the initial reaction of the acid with the calcite in the rock sample, causing the calcite to dissolve. As the acid is consumed, the pH of the

Table 3.4 Acid pH results at different concentrations

Hydrochloric acid concentration (%)	5/7	15	20	28	33
pH	– 0.3	– 0.6	– 0.6	– 1.3	– 1.6

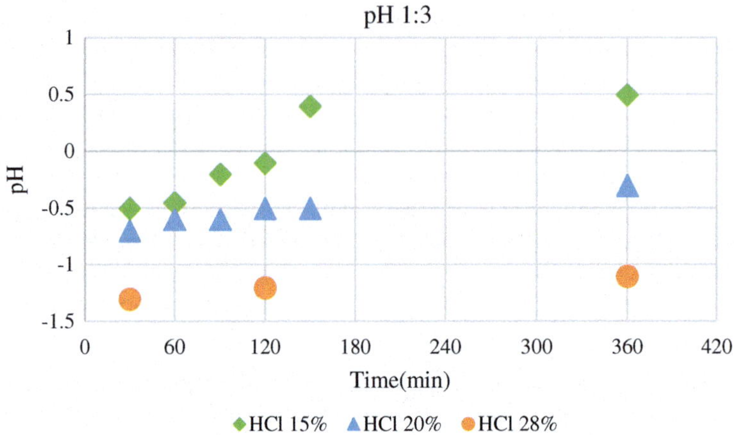

Fig. 3.13 Changes in pH over time for a 1:3 ratio

acidic medium increases, indicating the ongoing reaction between the rock and the acid. When the pH stabilizes and no longer changes, it signifies the end of the rock and acid reaction. For acids in this form, after about 120 min, the pH changes are minimal, indicating that the reaction is nearing completion. Among the three cases, 15% hydrochloric acid shows more significant pH changes. This is because 15% hydrochloric acid is weaker than the other two acids, so it reacts with the rock more quickly, leading to an earlier increase in pH.

Figures 3.14 and 3.15, the pH changes over time for hydrochloric acid with concentrations of 7.5, 15, 20, and 28% are plotted at a constant volume ratio of rock to acid, 1:5 and 1:7. Figure 3.13 clearly illustrates the initial reaction of the acid with the calcite in the rock sample, causing the calcite to dissolve. As the acid is consumed, the pH of the acidic medium increases, indicating that the rock-acid reaction is continuing. When the pH stabilizes and stops changing, this indicates the end of the rock-acid reaction. For acids in this form, the pH changes are minimal after about 120 min, indicating that the reaction is nearing completion. Of the three cases, 15% hydrochloric acid shows more significant pH changes. This is because 15% hydrochloric acid is weaker than the other two acids, so it reacts more quickly with the rock, leading to an earlier increase in pH. Figures 3.14 and 3.15 show the pH changes over time for hydrochloric acid at concentrations of 7.5, 15, 20, and 28% for constant rock-to-acid volume ratios of 1:5 and 1:7.

Figures 3.14 and 3.15 also show that as the acid concentration increases, its acid strength increases, resulting in a slower decrease in pH. In addition, the pH changes for 7.5% hydrochloric acid indicate that this acid is consumed more rapidly. As can be seen from Figs. 3.13, 3.14 and 3.15, the pH of the solution with a concentration of 28% falls into a lower range compared to the previous acids, mainly due to the high acid strength of this particular acid. Figure 3.16 provides a visualization of the pH variations for hydrochloric acid at different concentrations and rock/acid ratios.

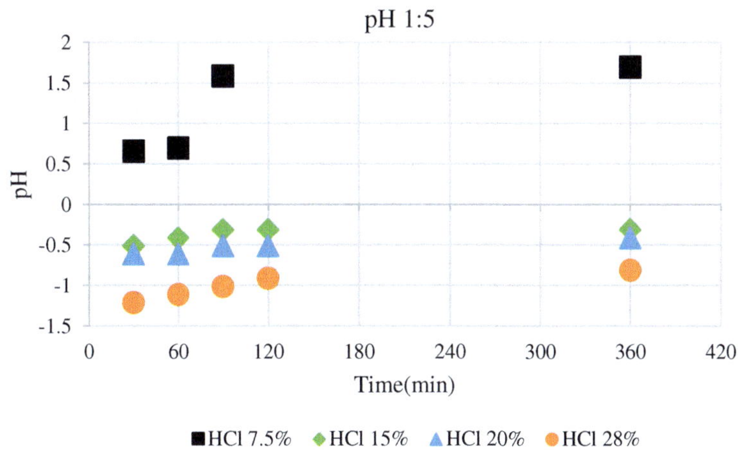

Fig. 3.14 Changes in pH over time for a 1:5 ratio

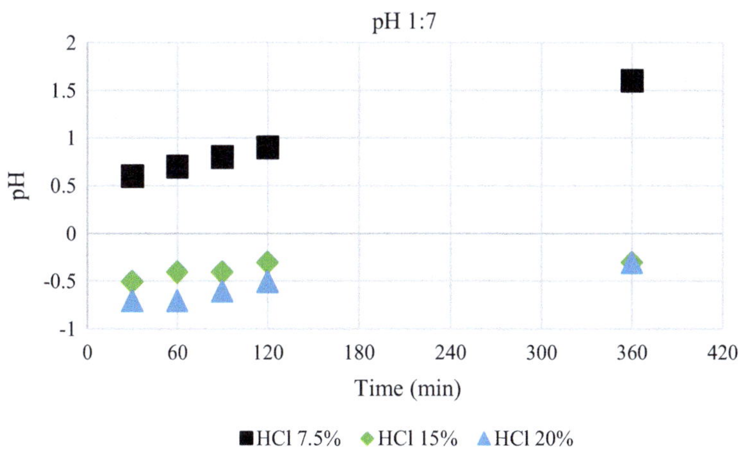

Fig. 3.15 Changes in pH over time for a 1:7 ratio

As shown in Fig. 3.16, an increase in acid concentration and volume relative to the rock results in a lower pH range, indicating the increased acid strength of the environment. In most cases, the pH remained unchanged after 100–150 min, indicating the end of the reaction.

Based on the dissolution and pH results, a rock-to-acid ratio of 1:7 was selected as the optimum volume for concentrations of 7.5, 15, and 20%, while a ratio of 1:5 was considered optimum for the 28% concentration. In addition, the period of 100–150 min marks the point at which acids are no longer able to dissolve rock.

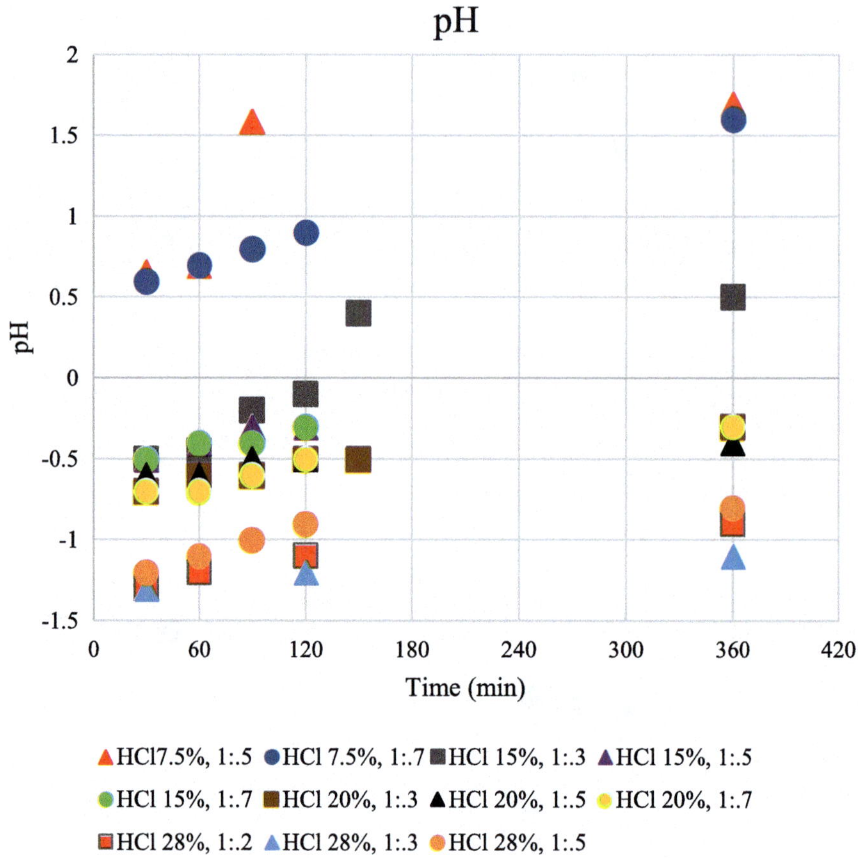

Fig. 3.16 Comparison of the pH of different acids in different volumes

3.5.2.3 Discussion About pH

Example 3.3 Explain the concept of pH and its significance in characterizing the acidity or basicity of a solution. How is the pH scale defined, and what values indicate acidic, neutral, and basic solutions?

Answer:

pH is a measure of the acidity or basicity of a solution, defined as the negative logarithm (base 10) of the hydrogen ion concentration ($[H_3O^+]$). The pH scale ranges from 0 to 14. Solutions with a pH less than 7 are acidic, pH 7 is neutral, and solutions with a pH greater than 7 are basic.

Example 3.4 Describe the relationship between the hydrogen ion concentration ($[H^+]$) and the pH of a solution. How does the concentration of hydrogen ions vary in acidic, neutral, and basic solutions?

Answer:

The pH of a solution is inversely proportional to the hydrogen ion concentration. As the concentration of hydrogen ions increases, the pH decreases, making the solution more acidic. In contrast, as the concentration decreases, the pH increases, indicating a more basic solution. Neutral solutions have equal concentrations of hydrogen and hydroxide ions, resulting in a pH of 7.

Example 3.5 Explain the effect of temperature on pH measurements. How does the pH of a solution change with temperature, and why is it important to consider temperature when interpreting pH values?

Answer:

Temperature can affect the ionization of water and influence pH measurements. As temperature increases, the ionization of water increases, leading to a higher concentration of hydrogen ions and a lower pH. It is crucial to consider temperature when interpreting pH values to ensure accurate assessments of a solution's acidity or basicity.

Example 3.6 What is a buffer solution, and how does it resist changes in pH? Describe the role of weak acids and their conjugate bases in maintaining the stability of buffer solutions.

Answer:

A buffer solution is a solution that resists changes in pH when an acid or base is added. It typically contains a weak acid and its conjugate base (or a weak base and its conjugate acid). The weak acid neutralizes added base, and the conjugate base neutralizes added acid, helping to maintain a stable pH.

Example 3.7 Discuss the relationship between pH and pOH. How are these values related, and what does their sum indicate about the nature of a solution?

Answer:

The relationship between pH and pOH is defined by the equation: $pH + pOH = 14$. If a solution has a low pH, it will have a high pOH, indicating a more basic solution. Conversely, a high pH corresponds to a low pOH, signifying a more acidic solution. A pH of 7 corresponds to a neutral solution where pOH is also 7.

Example 3.8 Explain the concept of acid–base titration. How does the addition of a base to an acid (or vice versa) during titration affect the pH of the solution? What key point on the titration curve represents the equivalence point?

Answer:

In acid–base titration, a base is gradually added to an acid (or vice versa), and the pH of the solution is measured. The equivalence point is reached when the moles of acid equal the moles of base. At this point, the solution is neutral, and the pH is close to 7.

3.5.3 Determining the Optimum Concentration

As mentioned in the previous sections, hydrochloric acid is used in concentrations ranging from 5 to 37%, with the most common concentration being 15%. Based on studies, HCl is classified into concentrations of 7.5, 15, 20, and 28%. In order to determine the optimum acid concentration, dissolution tests were carried out for different concentrations over a period of time. Consequently, the dissolution test aimed to estimate the calcite solubility for each concentration over a 24-h period at reservoir temperature (100 °C). Figures 3.17, 3.18, 3.19 and 3.20 show the stages before and after the solubility test, while Fig. 3.21 shows the solubility of calcite at different acid concentrations.

As can be seen in Fig. 3.21, the solubility of calcite increases with higher acid concentrations. According to this figure, concentrations of 20 and 28% are approximately 2–3% more soluble than the 15% concentration. In addition, concentrations above 17% show a reduced effect of inhibitors. Therefore, from both an economic and technical point of view, the choice of a lower concentration of acid can significantly improve the acidification process. Consequently, 15% was chosen as one of the optimal concentrations.

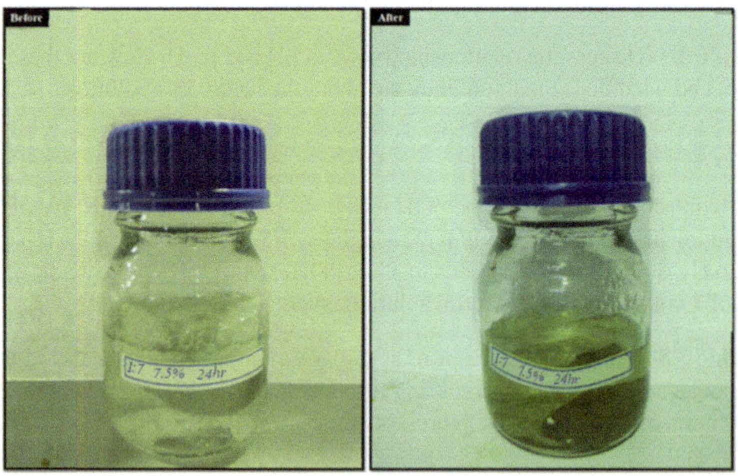

Fig. 3.17 Before and after placing the thin section of the plug in 7.5% HCl acid at a ratio of 1:7 for 24 h

Fig. 3.18 Before and after placing the thin section of the plug in 15% HCl acid at a ratio of 1:7 for 24 h

Fig. 3.19 Before and after placing the thin section of the plug in 20% HCl acid at a ratio of 1:7 for 24 h

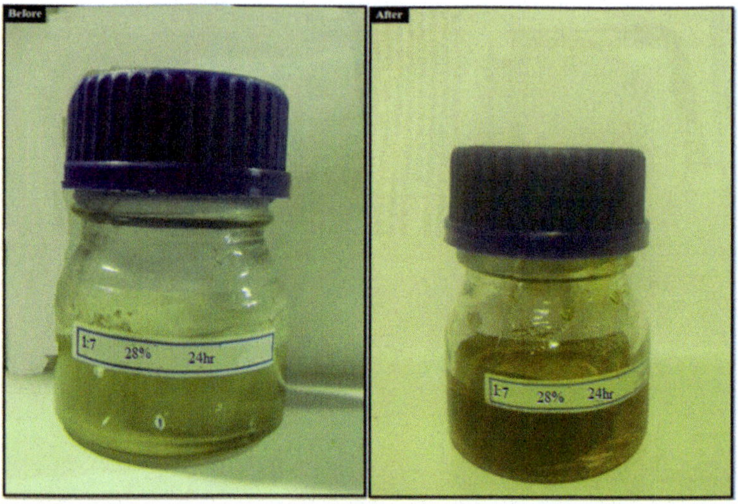

Fig. 3.20 Before and after placing the thin section of the plug in 28% HCl acid at a ratio of 1:5 for 24 h

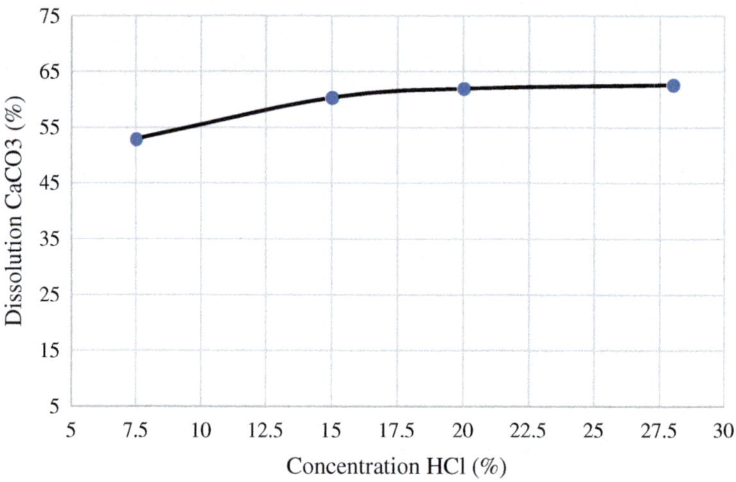

Fig. 3.21 Solubility rate of calcite for different concentrations of HCl

Given the low permeability of the Sarajeh field, weaker acids are advantageous as they have lower solubility at the wellhead. This allows the acid to penetrate the formation and increase the depth of acidizing. Therefore, 7.5% hydrochloric acid was selected as the second optimum concentration to simulate and test the operation with a weaker acid.

3.6 Hydrochloric Acid and Methanol

Methanol, as an alcohol, effectively reduces the surface tension of the solution, allowing greater penetration of the acid into smaller diameter pores. This allows the acid to access pores with high capillary pressure, ultimately improving dissolution. Capillary pressure depends on two key parameters: surface tension and capillary radius. As the capillary radius of pores remains relatively constant, reducing the surface tension is critical to overcoming the high capillary pressure of these pores and allowing the acid to enter smaller diameter pores.

Surfactants are commonly used to reduce the surface tension and achieve deeper penetration. Methanol is one such alcohol used alongside hydrochloric acid. Studies have shown that the use of methanol above 20% by volume with acid results in a reduction in surface tension, although this reduction is not significant. However, the use of methanol in concentrations above 20% poses a safety risk. Therefore, the recommended minimum volume percentage of methanol to be used with hydrochloric acid is 20% (Fig. 3.22).

3.6.1 Dissolution Test

Based on the information presented in the previous section, two acid combinations, namely 15% HCl + MeOH and 7.5% HCl + MeOH, with a volume ratio of 80/20, were subjected to dissolution tests for acidising operations. The results of the

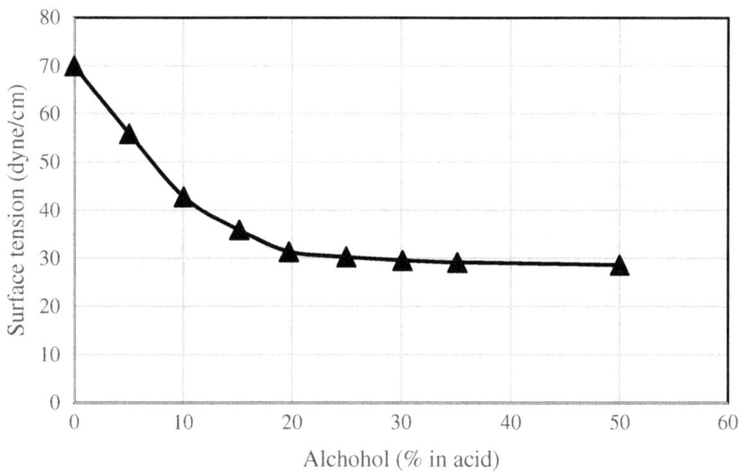

Fig. 3.22 Change of surface tension in different concentrations of methanol (Ali et al., 2016; Kalatehno et al., 2023)

experimental design for the dissolution test of these two acid mixtures are given in Table 3.5.

Table 3.6 shows the design algorithm of the hydrochloric acid and methanol test.

In this part, the dissolution test is similar to the previous case. The images obtained from these tests are shown in Figs. 3.23, 3.24, 3.25, 3.26, 3.27, 3.28, and 3.29.

Table 3.5 Hydrochloric acid and methanol test design

Time (min)	Concentration (%)	The ratio of rock to hydrochloric acid	The ratio of hydrochloric acid to methanol (%)
30	100% MeOH + 15% HCl	1:7	(20)MeOH – (80) HCl
90			
120	100% MeOH + 5/7% HCl		
360			

Table 3.6 Hydrochloric acid and methanol test design algorithm

Test number	Concentration (%)	Time (min)	Dissolution (%)	pH
1	HCl 7/5% + MeOH 100%	90	6/78	– 0.1
2	HCl 15% + MeOH 100%	360	39/78	– 0.5
3	HCl 15% + MeOH 100%	30	96/62	– 0.7
4	HCl 7/5% + MeOH 100%	360	39/78	0.1
5	HCl 7/5% + MeOH 100%	120	39/80	– 0.1
6	HCl 15% + MeOH 100%	120	89/76	– 0.6
7	HCl 7/5% + MeOH 100%	30	96/62	– 0.2
8	HCl 15% + MeOH 100%	90	31/66	– 0.6

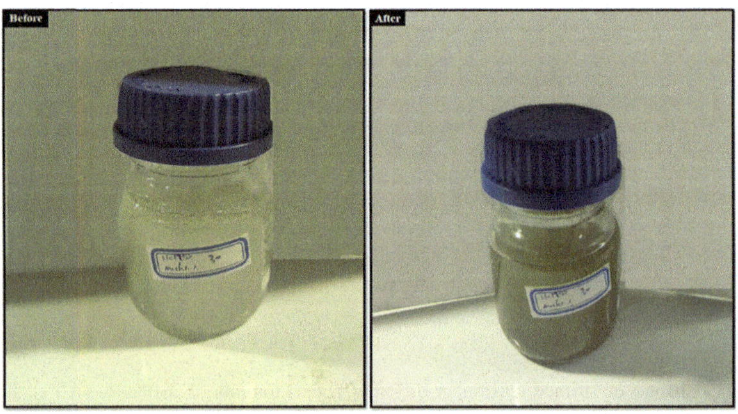

Fig. 3.23 Before and after placing the thin section of the plug in 7.5% (80) + MeOH (20)% HCl acid for 30 min

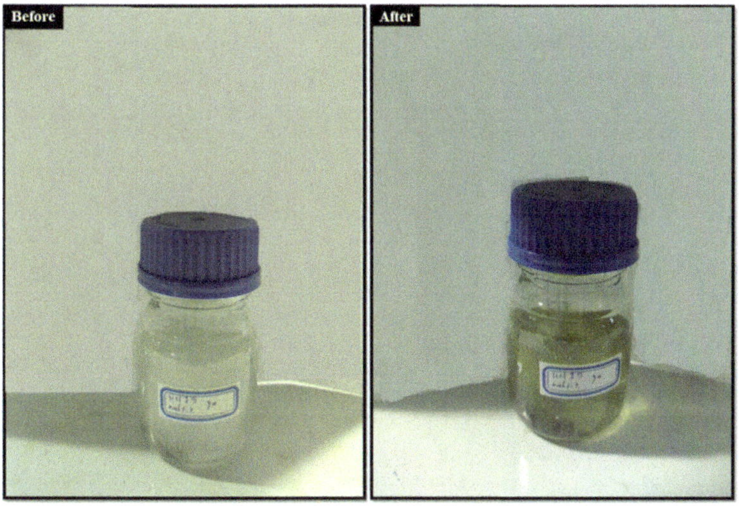

Fig. 3.24 Before and after placing the thin section of the plug in 7.5% (80) + MeOH (20)% HCl acid for 90 min

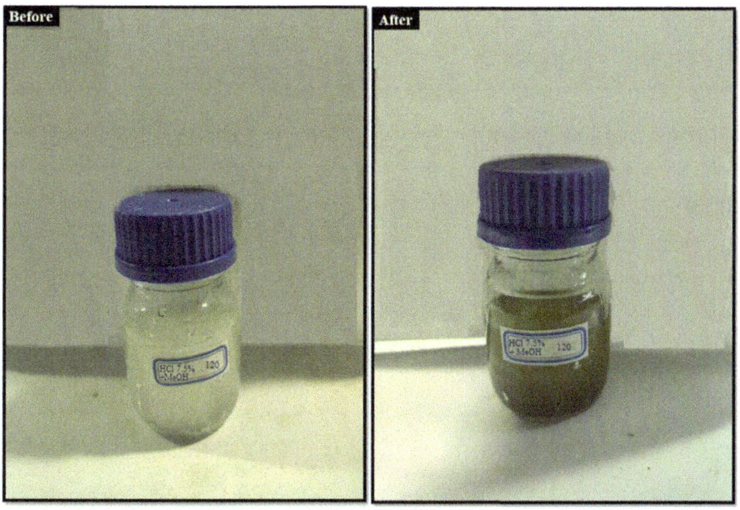

Fig. 3.25 Before and after placing the thin section of the plug in 7.5% (80) + MeOH (20)% HCl acid for 120 min

Fig. 3.26 Before and after placing the thin section of the plug in 15% (80) + MeOH (20)% HCl acid for 30 min

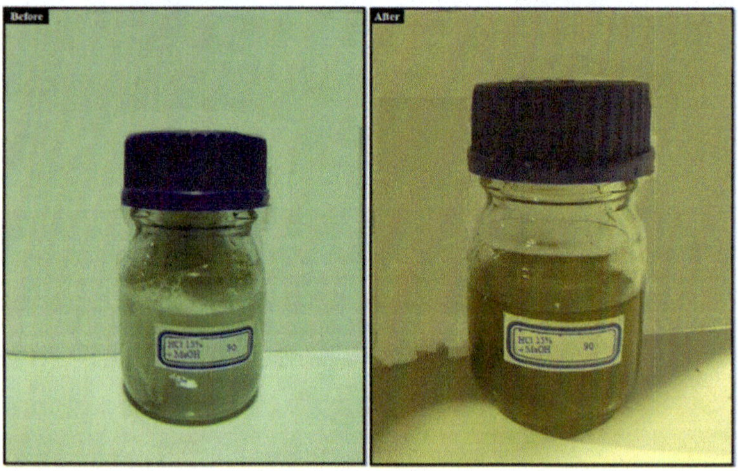

Fig. 3.27 Before and after placing the thin section of the plug in 15% (80) + MeOH (20)% HCl acid for 90 min

Fig. 3.28 Before and after placing the thin section of the plug in 15% (80) + MeOH (20)% HCl acid for 120 min

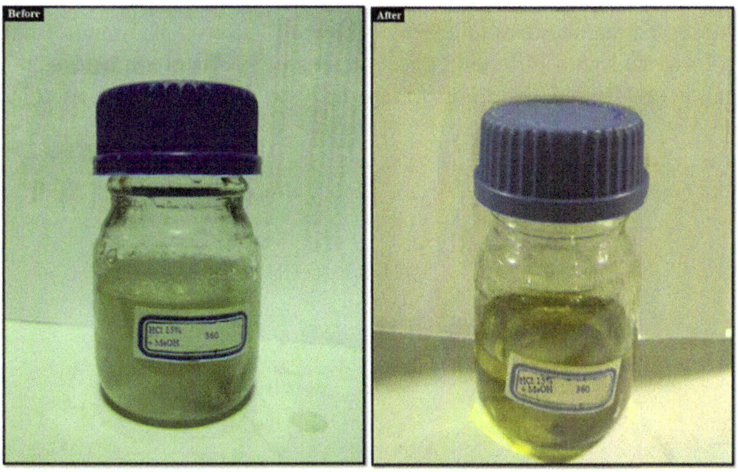

Fig. 3.29 Before and after placing the thin section of the plug in 15% (80) + MeOH (20)% HCl acid for 360 min

Figure 3.30 illustrates the percentage of calcite dissolution in a rock sample over time when exposed to a mixture of methanol and hydrochloric acid.

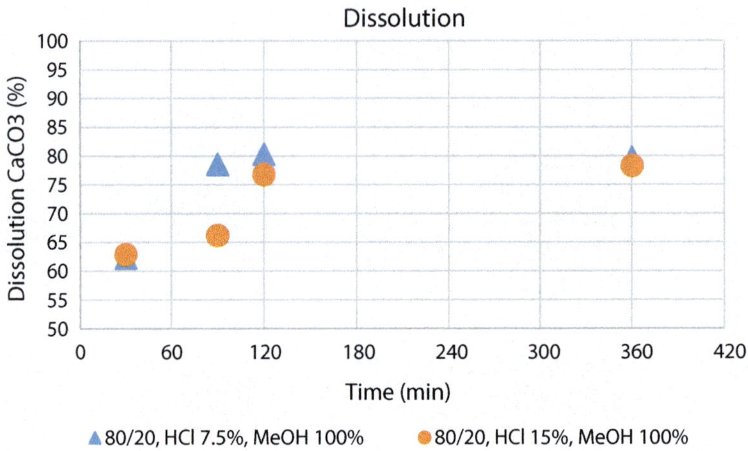

Fig. 3.30 Changes in the percentage dissolution of hydrochloric acid and methanol solution over time

3.6.1.1 pH

It was observed that the dissolution rates of two solutions containing methanol and hydrochloric acid with different concentrations of hydrochloric acid were almost identical. In this section, the pH results are analysed to determine the end time of the reaction. Figure 3.31 shows the pH plot for two solutions of hydrochloric acid and methanol. The orange circular dots represent 15% hydrochloric acid with methanol, while the blue triangular dots represent 7.5% hydrochloric acid with methanol.

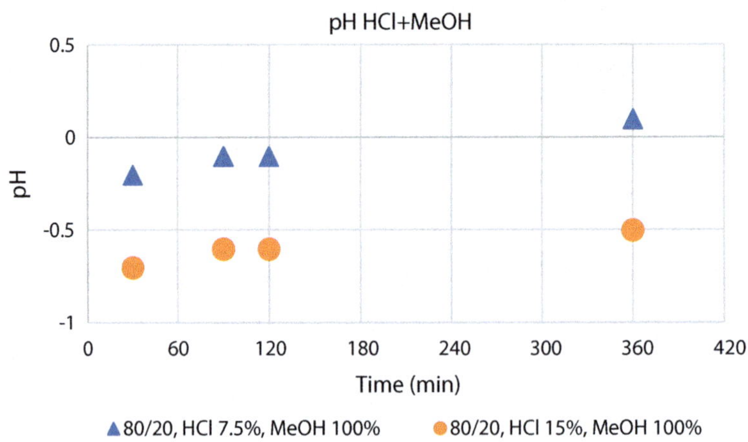

Fig. 3.31 pH changes of hydrochloric acid and methanol solutions

The pH of the 15% hydrochloric acid solution falls into a more acidic range during the reaction than the 7.5% hydrochloric acid solution. As a result, the 15% hydrochloric acid solution is more likely to react with rock in distant areas, affecting a larger portion of the environment. Therefore, the use of a 15% hydrochloric acid solution with methanol is expected to give better and more favorable results. As shown in Fig. 3.31, because 15% hydrochloric acid is a stronger acid than 7.5% hydrochloric acid, the pH range is shifted to lower values. In addition, the pH changes occur over a very small range and remain relatively constant, so the pH can be considered to be almost constant. It's worth noting that variations in ambient temperature could introduce some error in the pH values, potentially leading to slightly higher or lower values than actual. In conclusion, both solutions have been included in the list of injectable acids.

3.7 13% Acetic Acid Dissolution Test

Mineral acids are always used as the fixed fluid in acidising wells. Hydrochloric acid is one of the mineral acids used in carbonate reservoirs. The acid strength of this type of acid is high; therefore, the use of only this acid leads to corrosion of the pipes. On the other hand, most of the acid volume is consumed in the same area near the well and the depth of penetration is not sufficient. The use of organic acids together with mineral acids not only reduces corrosion but also slows down the reaction rate of rock and acid, allowing the acid to penetrate further and affect a larger area. The results of the experimental design for the 13% acetic acid dissolution test are given in Table 3.7 (Da Motta et al., 1996, 1998; Harris, 1961; Huang et al., 2000).

The extent of calcite dissolution was measured at 60, 120, 240, 360, and 1440 min for 13% acetic acid. This part of the dissolution test follows a similar procedure to the previous case. The images taken during these experiments are shown in Figs. 3.32, 3.33, 3.34, 3.35, and 3.36. (Table 3.8).

The dissolution rate of calcite in the presence of 13% acetic acid is shown in Fig. 3.37.

Table 3.7 Acetic acid test design

Acid type	Acid volume ratio	Rock to acid ratio	Concentration (%)	Time (min)
Acetic acid	100	1:30	13%	60
				120
				240
				360
				1440

Fig. 3.32 Before and after placing the thin section of the plug in 13% HAc acid for 60 min

Fig. 3.33 Before and after placing the thin section of the plug in 13% HAc acid for 120 min

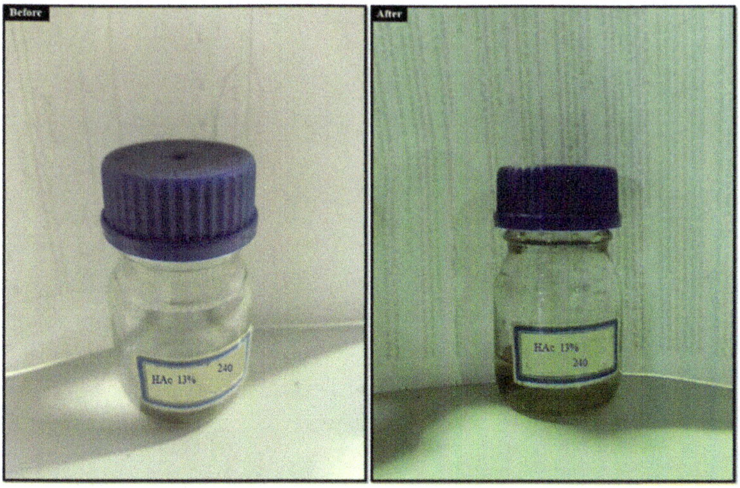

Fig. 3.34 Before and after placing the thin section of the plug in 13% HAc acid for 240 min

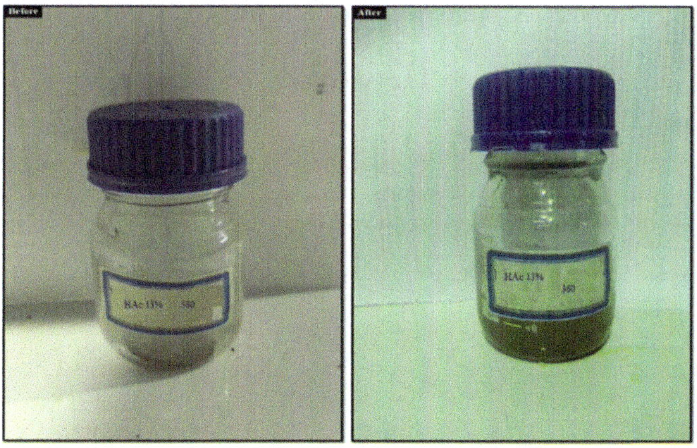

Fig. 3.35 Before and after placing the thin section of the plug in 13% HAc acid for 360 min

As shown in Fig. 3.37, the initial calcite dissolution rate is approximately 22–23%. After 6 h this rate increases to 28%. It's worth noting that the calcite content in this sample is about 15% higher than in the other samples. After 24 h of reaction, the calcite dissolution rate stabilizes at 20%. The slight decrease in solubility observed here could be attributed to the formation of calcium acetate precipitate, which inhibits further acid-rock reactions. Figure 3.38 shows the rock sample before the acid reaction, while Fig. 3.39 shows the calcium acetate precipitate formed after 24 h of testing (Qiu & Chang, 2011).

Fig. 3.36 Before and after placing the thin section of the plug in 13% HAc acid for 1440 min

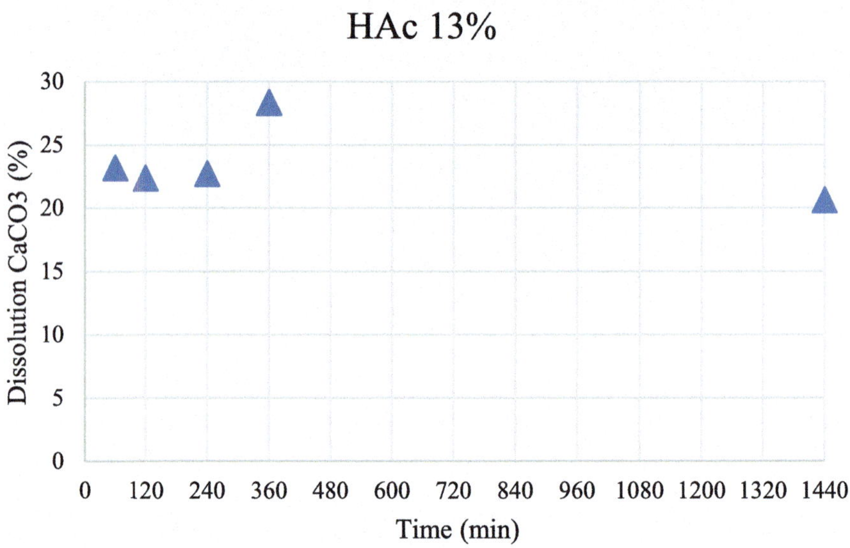

Fig. 3.37 The percentage of calcite dissolution in the presence of acetic acid is 13%

Fig. 3.38 Rock sample before reaction with acetic acid

Fig. 3.39 Rock sample after reaction with acetic acid

Table 3.8 Design algorithm of acetic acid test

Test number	Concentration (%)	Time (min)	Dissolution	pH
1	HAc 13%	60	23.19	4.3
2	HAc 13%	120	22.36	4.6
3	HAc 13%	240	22.71	4.7
4	HAc 13%	360	28.34	5.1
5	HAc 13%	1440	20.62	5.3

In addition to all the other factors, checking the density results can also be useful in checking the dissolution of the rock. Table 3.9 shows the measured density after the reaction of 13% acetic acid with rock within 24 h.

As can be seen in Table 3.9, the density of the solution increased with time, except at 1440 min, which is a sign of the dissolution of the rock by the acid. At 1440 min the density has decreased and this decrease in density may be due to the precipitation of calcium acetate which caused part of the solution to crystallize and come out of the solution, reducing the density.

Table 3.9 Density of 13% acid before and after reaction

Time (min)	Density before reaction (g/cm^3)	Density after reaction (g/cm^3)
60	1.0054	1.19
120	1.0054	1.20127
240	1.0054	1.23
360	1.0054	1.2342
1440	1.0054	1.12316

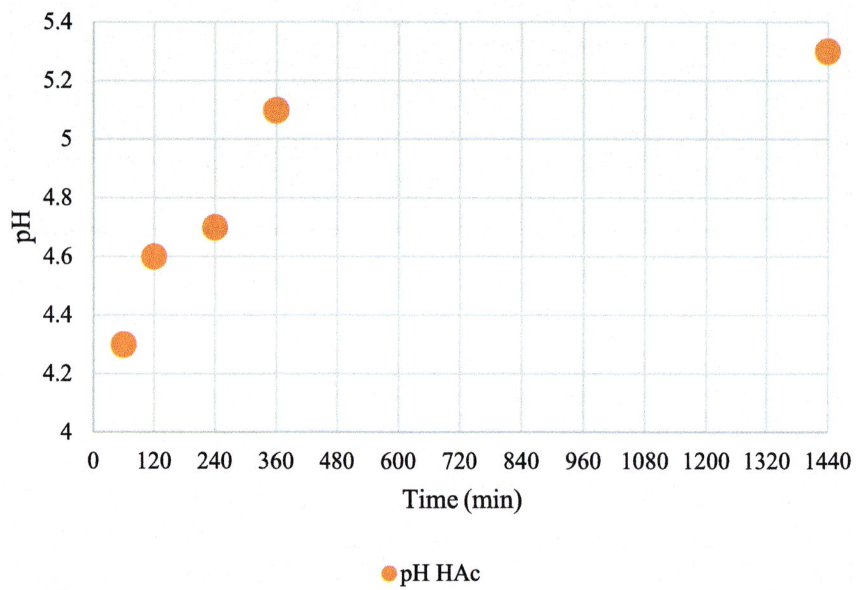

Fig. 3.40 pH changes for 13% acetic acid

3.7.1 *Acetic Acid pH at 13%*

Figure 3.40 shows the pH of 13% acetic acid over time.

In this graph, the pH shows an upward trend and remains relatively stable at 1440 min. The rising pH indicates the ongoing reaction between the rock and the acid. When there is no further change in pH, it signals the completion of the reaction. The pH shifts from 4.3 to 5.2 over time. The minimal changes in pH indicate that acetic acid is much weaker than hydrochloric acid. In addition, the initial pH of 13% acetic acid is 2.8 before the reaction starts. It is clear that the use of acetic acid alone in acidification may not give satisfactory results. It is more effective when combined with mineral acids to achieve a greater depth of penetration. The reaction rate of acetic acid is slower than that of hydrochloric acid. Therefore, when used in conjunction with hydrochloric acid, it facilitates the distribution of the acidic solution, thus achieving a greater area of action.

3.8 Determination of the Optimum Time and Volume Ratio of Hydrochloric Acid to Acetic Acid

3.8.1 *Dissolution Test*

Acetic acid is typically used in concentrations of 2.5, 8, 10, and 13%, often in combination with hydrochloric acid. The 8 and 13% concentrations are often combined with 28 and 15% hydrochloric acid. In particular, the combination of 13% acetic acid with 15% hydrochloric acid has a higher solubility than that of 8% acetic acid with 15% hydrochloric acid. In addition, according to the studies carried out, 7.5% hydrochloric acid is used in addition to 2.5% acetic acid.

The results of the experimental design to investigate the dissolution test of 13% acetic acid and hydrochloric acid with volume ratios of 75/25, 50/50, and 25/75 are presented in Tables 3.10 and 3.11. The use of different ratios in this experiment is driven by the objective of optimizing cost, reaction rate, and the ratio of acetic acid to hydrochloric acid. In this section the dissolution test follows a similar procedure to the previous cases and images of these tests are shown in Figs. 3.41, 3.42, 3.43, 3.44, and 3.45.

Table 3.11 shows the design algorithm for the hydrochloric acid and acetic acid tests.

Figures 3.46, 3.47, and 3.48 show the comparison of the rate of dissolution of 13% acetic acid with the acid solution of 7.5% HCl and 2.5% HAc in different ratios.

As can be seen from all three graphs, the optimum ratio of acetic acid to hydrochloric acid was found to be 25/75, as shown in Fig. 3.46, taking into account the maximum dissolution. Figure 3.45 shows that the maximum dissolution obtained

Table 3.10 Hydrochloric acid and acetic acid test design

The ratio of acetic acid to hydrochloric acid	Hydrochloric acid rock-to-acid ratio	Concentration	Time (min)
75 HAc/25 HCl	1:7	13% HAc + 15% HCl 2.5% HAc + 7.5% HCl	30
50 HAc/50 HCl			90
25 HAc/75 HCl			120
			360

with 13% acetic acid is slightly lower than that obtained with a combination of 7.5% HCl and 2.5% HAc. However, in the other two figures, the dissolution rate of 13% acetic acid is almost identical to that achieved with 7.5% HCl and 2.5% HAc, with only a small percentage difference. Therefore, the choice between these two acids at low ratios does not significantly affect the solubility. It can therefore be concluded that a higher proportion of hydrochloric acid in the acidic solution leads to increased solubility. Figures 3.49, 3.50, and 3.51 show a comparison of the dissolution rates for 13% acetic acid with 15% HCl acid solution and 13% HAc at different ratios.

As can be seen from all three graphs, the optimum ratio of acetic acid to hydrochloric acid was determined to be 75/25, taking into account the greater dissolution observed in Fig. 3.49. Figure 3.49 clearly shows that the maximum dissolution of acetic acid at 13% is slightly lower than that achieved with 15% HCl and 13% HAc. It is therefore evident that a higher proportion of hydrochloric acid in the acidic solution results in greater solubility. Next, to determine the optimum acid, a comparison is made between two acid solutions: 15% HCl and 13% HAc, and 7.5% HCl and 2.5% HAc, both with an acetic acid/hydrochloric acid ratio of 75/25. Figure 3.52 shows the comparison of the dissolution rates of these two solutions.

As shown in the graph, the average solubilities of these two acids are very similar. After 360 min, solutions containing 7.5% HCl and 2.5% HAc have a higher solubility.

3.8.2 pH

Figure 3.53 shows the pH changes for two samples of hydrochloric acid and acetic acid with a ratio of 75/25 acetic acid to hydrochloric acid. According to the solubility results, the ratio of 25/75 acetic acid to hydrochloric acid was chosen as the optimum ratio, so only the graph relating to this ratio was analyzed.

The combination of 7.5% hydrochloric acid and 2.5% acetic acid is a much weaker acid than the mixture of 15% hydrochloric acid and 13% acetic acid. However, in the solubility section, it was observed that these two acids have almost identical solubilities. As the reaction begins, the acids are consumed, leading to an increase in pH. Specifically, the 7.5% hydrochloric acid and 2.5% acetic acid mixture, being

Table 3.11 Hydrochloric acid and acetic acid test design algorithm

Test number	The ratio of acetic acid to hydrochloric acid	Concentration (%)	Time (min)	Dissolution (%)	pH
1	50% HCl + 50% HAc	HCl 15% + HAc 13%	120	43.95	0.5
2	25% HCl + 75% HAc	HCl 7.5% + HAc 2.5%	360	38.7	3.7
3	75% HCl + 25% HAc	HCl 15% + HAc 13%	90	50.95	−0.2
4	25% HCl + 75% HAc	HCl 7.5% + HAc 2.5%	120	27.71	1.9
5	75% HCl + 25% HAc	HCl 15% + HAc 13%	120	47.03	−0.2
6	75% HCl + 25% HAc	HCl 7.5% + HAc 2.5%	90	33.16	0.2
7	25% HCl + 75% HAc	HCl 7.5% + HAc 2.5%	30	11.55	0.9
8	25% HCl + 75% HAc	HCl 15% + HAc 13%	120	25.73	3.4
9	25% HCl + 75% HAc	HCl 15% + HAc 13%	30	17.73	0.4
10	75% HCl + 25% HAc	HCl 7.5% + HAc 2.5%	120	51.42	0.3
11	50% HCl + 50% HAc	HCl 15% + HAc 13%	30	28.27	0.1
12	75% HCl + 25% HAc	HCl 7.5% + HAc 2.5%	30	34.88	0.1
14	50% HCl + 50% HAc	HCl 7.5% + HAc 2.5%	120	40.38	0.9
15	50% HCl + 50% HAc	HCl 15% + HAc 13%	360	39.96	1.4
16	50% HCl + 50% HAc	HCl 15% + HAc 13%	90	43.95	0.5
17	50% HCl + 50% HAc	HCl 7.5% + HAc 2.5%	30	14.9	0.4
18	50% HCl + 50% HAc	HCl 7.5% + HAc 2.5%	90	31.31	0.6
19	75% HCl + 25% HAc	HCl 7.5% + HAc 2.5%	360	59.27	0.4
20	25% HCl + 75% HAc	HCl 15% + HAc 13%	90	27.19	2/2
21	25% HCl + 75% HAc	HCl 7.5% + HAc 2.5%	90	20.43	1/1
22	75% HCl + 25% HAc	HCl 15% + HAc 13%	30	34.71	−0.3

(continued)

Table 3.11 (continued)

Test number	The ratio of acetic acid to hydrochloric acid	Concentration (%)	Time (min)	Dissolution (%)	pH
23	25% HCl + 75% HAc	HCl 15% + HAc 13%	360	46.2	3.7
24	50% HCl + 50% HAc	HCl 7.5% + HAc 2.5%	360	43.82	1/1

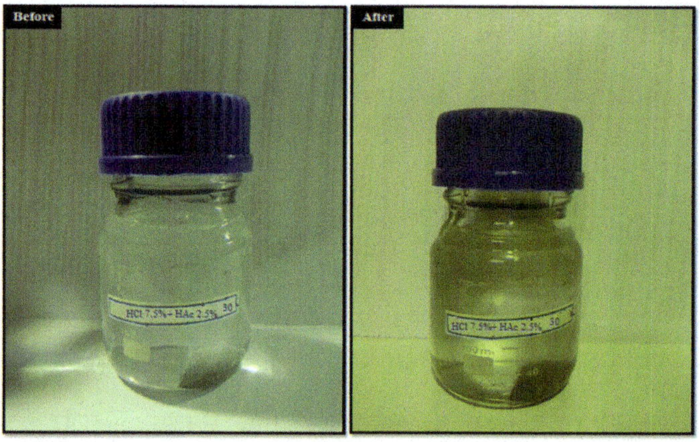

Fig. 3.41 Before and after placing the thin section of the plug in HCl 7.5% (25) – HAc 2.5% (75) acid for 30 min

the weaker acid, results in a higher pH compared to the 15% hydrochloric acid and 13% acetic acid combination. The main contributor to the lower pH is the stronger hydrochloric acid. Consequently, when a stronger hydrochloric acid is used, the pH tends to fall into a lower range.

By increasing the volume of acetic acid relative to hydrochloric acid, the pH of the reaction solution will change in the range of 0–4. Ultimately, due to the near identical solubility of these two solutions, the combination of HCl 7.5% and HAc 2.5%, with an acetic acid to hydrochloric acid ratio of 75/25, emerges as the optimal acid candidate in this context. This choice is driven by its cost-effectiveness compared to the 15% HCl and 13% HAc solution, as well as its ability to improve penetration depth.

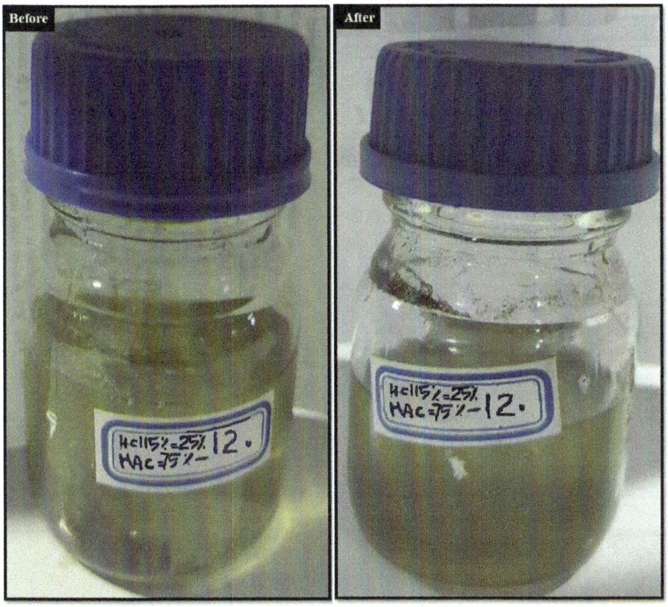

Fig. 3.42 Before and after placing the thin section of the plug in HCl 15% acid (25) – HAc 13% (75) for 120 min

Fig. 3.43 Before and after placing the thin section of the plug in HCl 15% (50) – HAc 13% (50) acid for 30 min

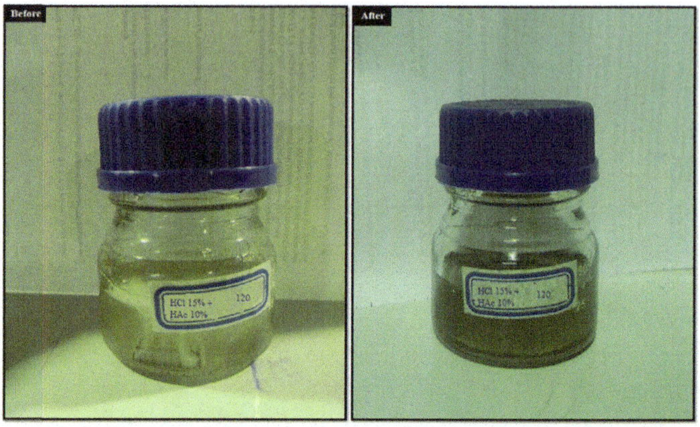

Fig. 3.44 Before and after placing the thin section of the plug in HCl 15% (50) – HAc 13% (50) acid for 120 min

Fig. 3.45 Before and after placing the thin section of the plug in HCl 7.5% (75) – HAc 2.5% (25) percent for 360 min

3.9 Viscosity Increasing Fluids

Viscosity-enhancing fluids, also known as polymer-viscosified acids or foams, have potential applications in all types of well completions. These fluids are particularly useful in horizontal and open-hole wells where many acid diversion techniques may not be feasible. Fluids that increase the viscosity of the acid create higher flow resistance in areas of high permeability. As a result, acid penetration extends into these areas as well as others. A more effective alternative to polymers or foams is the use of gel acids. Gel acid, being a non-Newtonian fluid, exhibits increased viscosity

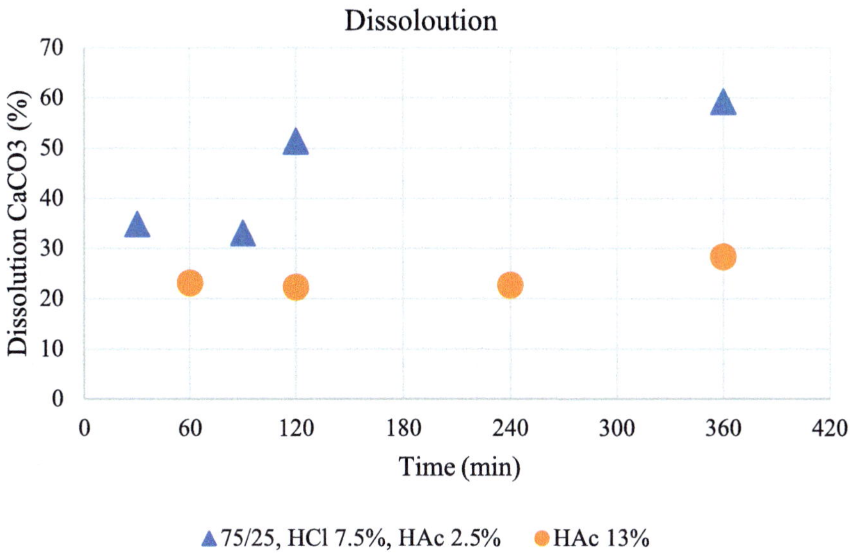

Fig. 3.46 Comparison of the dissolution rate of HAc 13% with HCl 7.5% + HAc 2.5% (25/75)

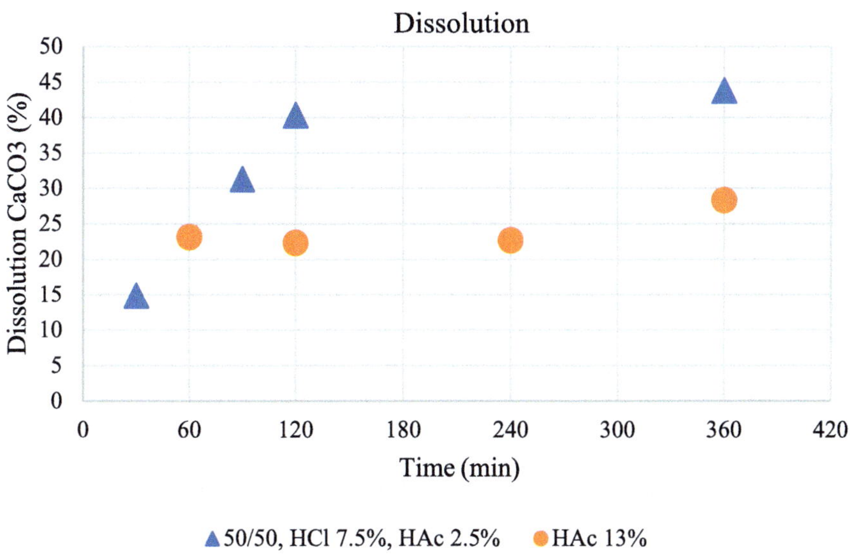

Fig. 3.47 Comparison of the dissolution rate of HAc 13% with HCl 7.5% + HAc 2.5% (50/50)

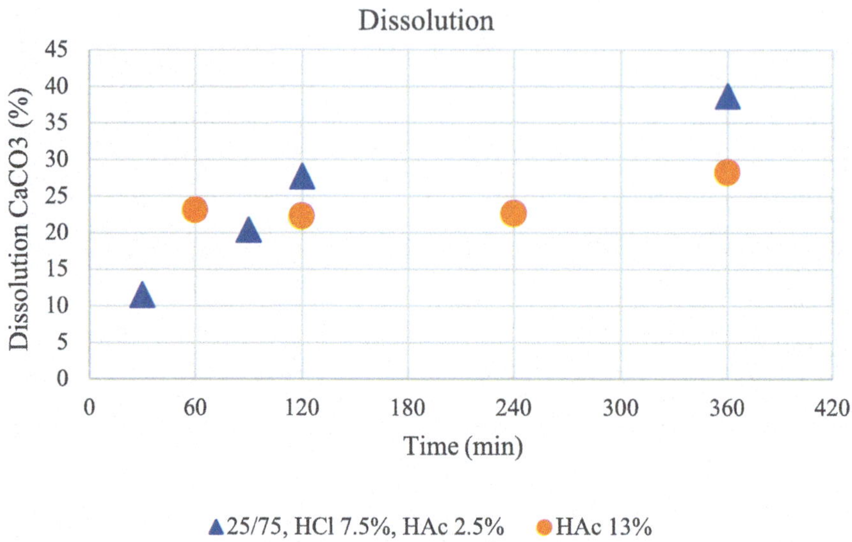

Fig. 3.48 Comparison of the dissolution rate of HAc 13% with HCl 7.5% + HAc 2.5% (75/25)

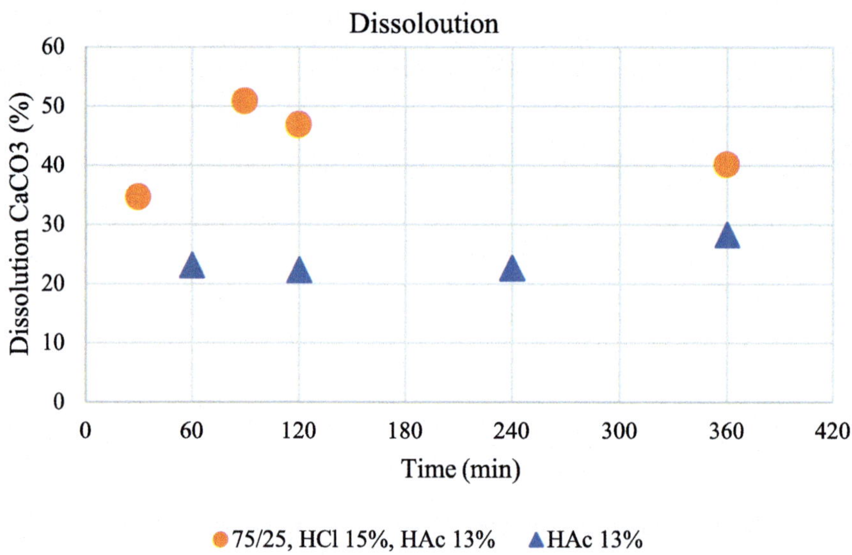

Fig. 3.49 Comparison of the dissolution rate of HAc 13% with HCl 15% HAc 13% (25/75)

as it travels further from the wellbore. This property effectively diverts the acid to areas of lower permeability. Other acid diversion systems proposed and used include acid combined with a gelling agent and a pH-sensitive crosslinking agent. During initial acid consumption within the wormhole, the pH rises to a range of 2–3. This

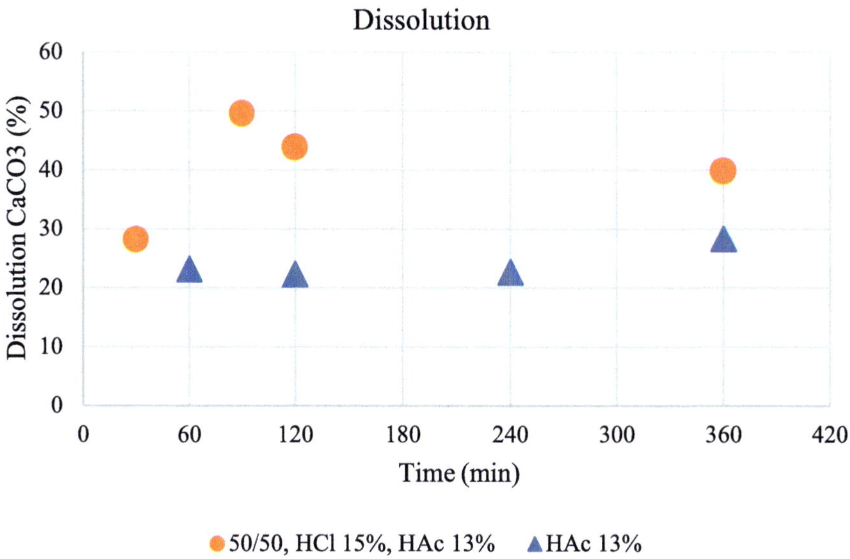

Fig. 3.50 Comparison of the dissolution rate of HAc 13% with HCl 15% + HAc 13% (50/50)

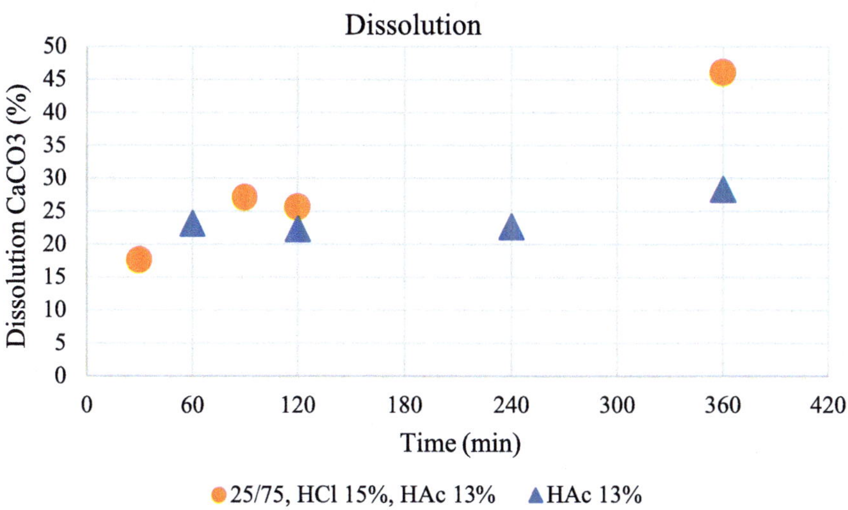

Fig. 3.51 Comparison of the dissolution rate of HAc 13% with HCl 15% HAc + 13% (25/75)

significant increase in pH leads to a corresponding increase in acid viscosity due to the action of the crosslinker. Subsequently, as more acid is consumed, the crosslinker breaks down. At a pH above 4, the viscosity of the fluid returns to its original state. Such a pH-sensitive crosslinked gel is suitable for long-range acid displacement as shown in Fig. 3.54.

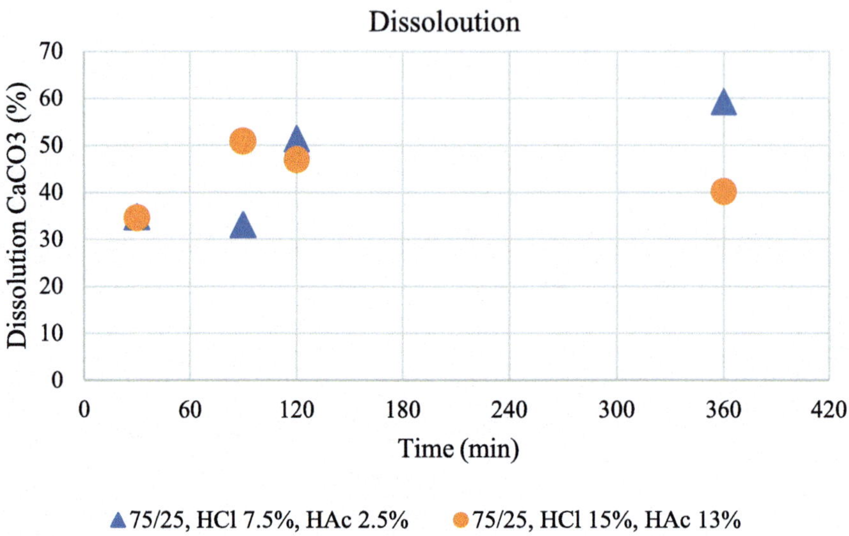

Fig. 3.52 The percentage of dissolution of hydrochloric acid and acetic acid solution in different concentrations in constant volume (75/25)

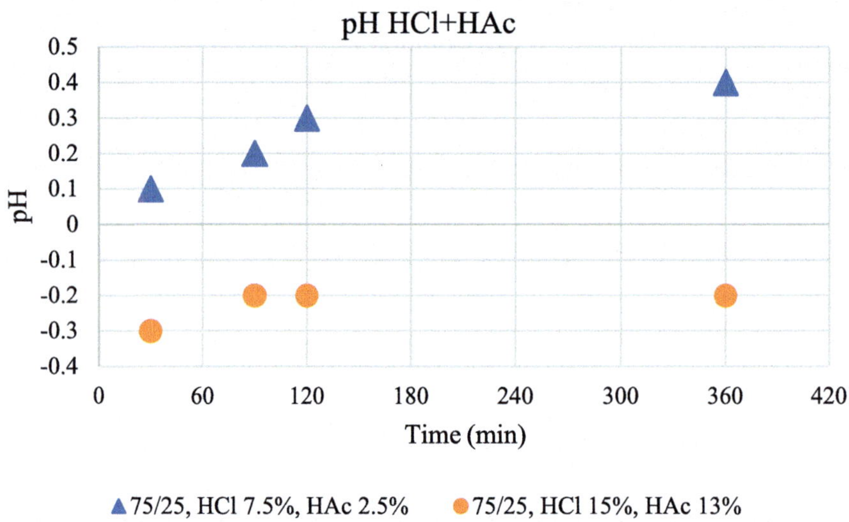

Fig. 3.53 Changes in pH with time for different concentrations of hydrochloric acid and acetic acid in constant volume (75/25)

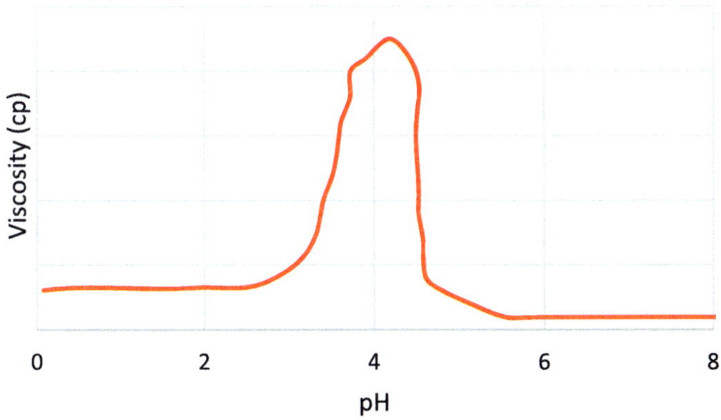

Fig. 3.54 The graph of gel acid viscosity changes with pH (Toseef et al. 2010; Ratnakar et al., 2012)

The study of low-permeability carbonate formations has shown that the application of gel acid and viscoelastic fluids results in the closure of existing pores. The injection of such fluids effectively disrupts the existing permeability, resulting in the cessation of production. Consequently, the use of this type of acid in formations with a permeability of around 1 millidarcy is not recommended.

3.10 Comparison of Selected Solutions

So far, 6 solutions have been selected, as shown in Table 3.12 In the following, these 6 solutions will be examined together in order to select the optimum acid.

Figures 3.55 and 3.56 illustrate the solubility of hydrochloric acid when combined with acetic acid and methanol.

These graphs show that when hydrochloric acid is used without methanol or acetic acid, its solubility is equal to that of hydrochloric acid combined with acetic acid. In particular, it's important to note that the combination of acetic acid and hydrochloric

Table 3.12 Selected acids

Acid type	The ratio of rock volume to hydrochloric acid	Volume ratio of hydrochloric acid to acetic acid/methanol
HCl 7.5%	1:7	–
HCl 7.5% + HAc 2.5%	1:7	75/25
HCl 7.5% + MeOH	1:7	80/20
HCl 15%	1:7	–
HCl 15% + HAc 10%	1:7	75/25
HCl 15% + MeOH	1:7	80/20

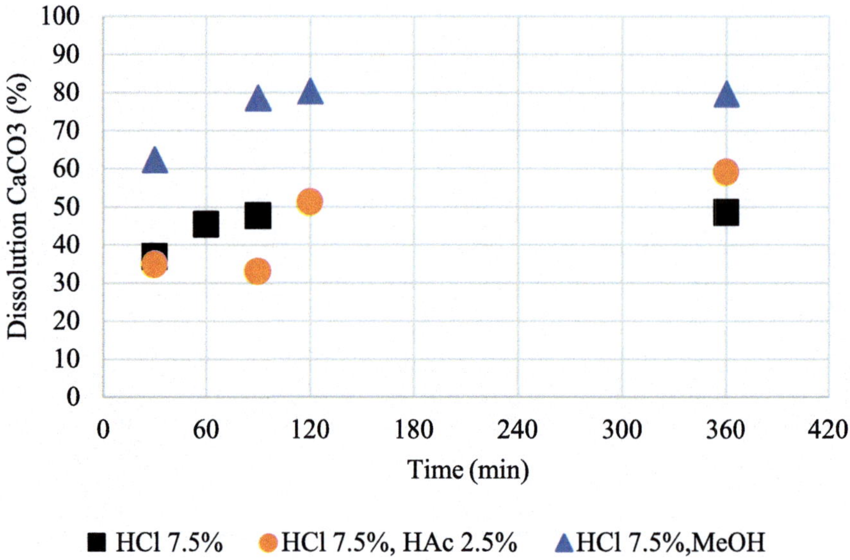

Fig. 3.55 Comparing the dissolution rate of 7.5% hydrochloric acid in different solutions

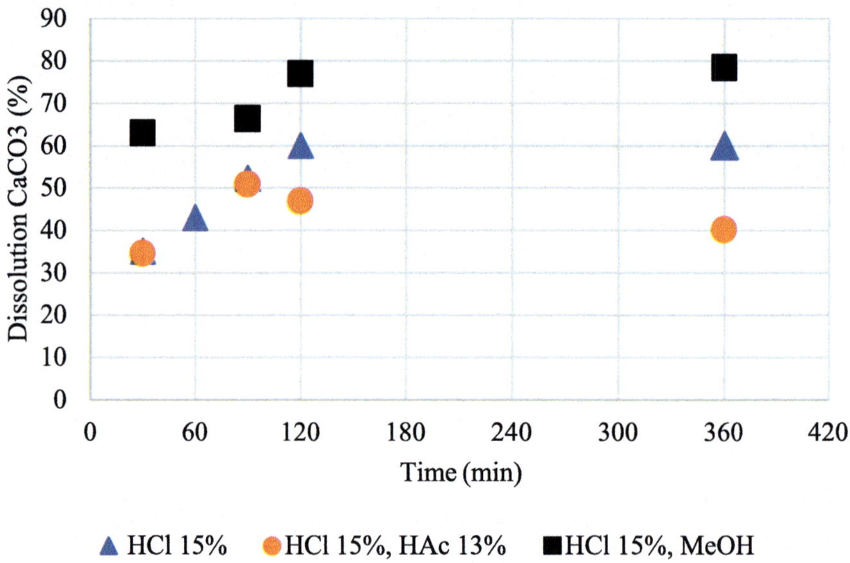

Fig. 3.56 Comparing the dissolution rate of 15% hydrochloric acid in different solutions

acid provides a greater depth of penetration, according to the results of the articles. On the other hand, the solution of hydrochloric acid with methanol not only achieves a greater depth of penetration but also shows a higher degree of solubility. Methanol allows deeper penetration into smaller pores by reducing the surface tension.

Therefore, according to Fig. 3.55, HCl 7.5% + MeOH and HCl 7.5% + HAc 2.5% and according to Fig. 3.56, HCl 15% and HCl 15% + MeOH have been selected as the optimum injection fluids. The 15% HCl + 13% HAc solution was not selected as the optimum acid due to its similar solubility to 15% HCl and higher cost compared to the 7.5% HCl + 2.5% HAc solution, as well as its weak acid and similar solubility to HCl 7.5% + 2.5% HAc.

3.11 Perform Surface Tension and Contact Angle Tests

3.11.1 Introduction

The pore volume within the reservoir rock is not limited to a single phase but includes gas, oil, and water saturations. As a result, we often encounter not only two but in many cases three phases in reservoirs. When the pore spaces of the reservoir rock contain only a single fluid, we need to consider only a single set of forces associated with the interaction between the rock and that specific fluid. However, when multiple fluid phases coexist, we need to consider three distinct forces: those between the first fluid and the rock, the second fluid and the rock, and the first fluid and the second fluid. The presence of these forces affects fundamental properties such as interfacial tension and wettability. In reservoir rocks where multiple fluids are present, a comprehensive understanding of these properties is essential to accurately characterize the potential of production and injection fluids. This chapter will focus on interfacial tension and wettability. The first set of forces to consider are interfacial or interfacial tension forces, as wettability is closely related to interfacial tension. In multiphase reservoirs, especially when these fluids are immiscible and come into contact with each other, surface tension phenomena occur between the phases.

3.11.2 Acidizing Additives

Choosing the right acid for acidizing is a critical decision. The addition of specific additives is necessary to achieve the optimum acid. These additives serve a variety of purposes, including the prevention of excessive corrosion, slug, emulsion formation, iron precipitation, and overall improvement of the desired acid properties. A constant additive in any acid process is the corrosion inhibitor. This addition effectively reduces the rate of corrosion of metal pipes and equipment. Non-emulsifiers play a vital role in preventing the formation of acid/hydrocarbon emulsions. In addition, surfactants lower the surface tension and change the wettability to hydrophilic, allowing deeper penetration and reaction of the rock with the acid. The presence of iron in the acidizing solution can lead to iron precipitation in the formation when the pH of the acid used reaches 2. To overcome this problem, an iron reducer is used to convert ferric to ferrous, which precipitates at pH 7, thus preventing unwanted iron precipitation. An iron controller is also used to keep ferric in solution and prevent its conversion to ferrous. In deep wells, where significant pressure drops occur along the wellbore during acidizing operations, it is necessary to minimize this pressure drop. This is achieved by using a friction reducer which allows the pump to operate at reduced power. In cases where the reaction rate between the rock and acid is exceptionally high, resulting in rapid acid consumption, the use of retarders is essential. Retarders are used to reduce the reaction rate, allowing the acid to penetrate the formation more effectively. In addition, the reaction rate between acid and rock is greatly accelerated in high-temperature wells. Hydrochloric acid in particular has a rapid reaction rate and consumption. To overcome this problem, retarders are used to slow down the reaction rate and ensure effective acid treatment.

Table 3.13 contains a list of the additives produced by the research group. This table lists the name of each additive, its photograph, and its abbreviated name.

3.11.2.1 Iron Control

As mentioned above, ferrous iron precipitates at a pH close to 7, while ferric iron precipitates at a pH close to 2. As the pH of our acid rarely reaches 6, ferrous iron is less of a problem for us. It's mainly ferric iron that contributes to most of our challenges due to its tendency to precipitate at lower pH levels. Iron can come from several sources, including:

- Corrosion products on the tube wall.
- Rust.
- Iron-bearing minerals.

Formation water and formation minerals that naturally contain iron. However, the most significant source of iron is usually the tubing inside the well. These pipes often contain Fe_2O_3, which can dissolve in the acid. The extent of dissolution depends on factors such as the duration of contact with the acid and temperature control. As this compound dissolves, it is converted to ferric iron, resulting in the formation of an iron hydroxide precipitate ($Fe(OH)_3$). This precipitate is gelatinous and brown and has the potential to damage the formation.

Table 3.13 Additives used in acid design

Additive name	Abbreviated name	Photo
Iron control	WA 450 N	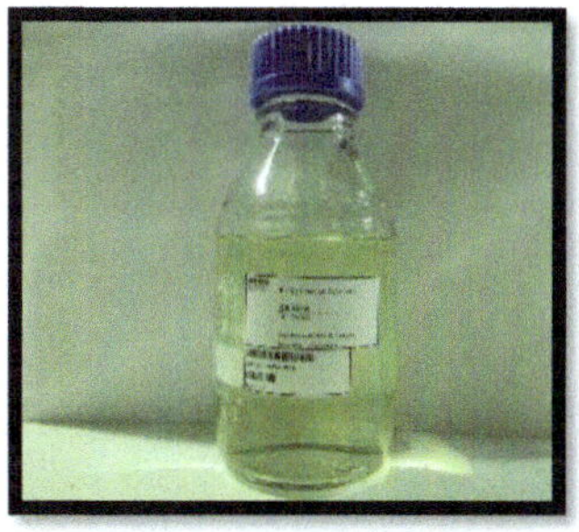
Iron reducer	WA IRED 457	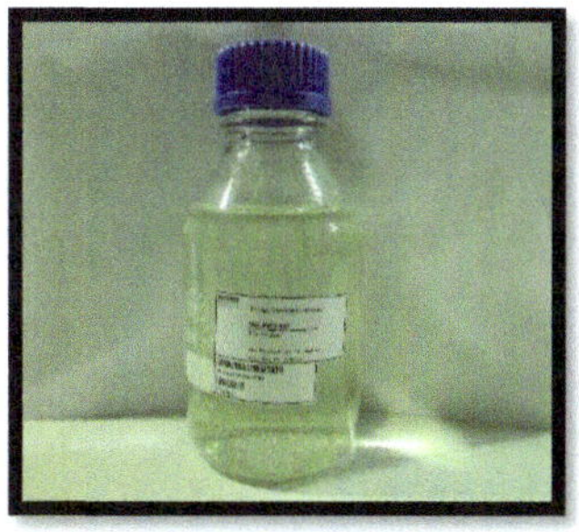
Frtiction reducer	WA FR 110	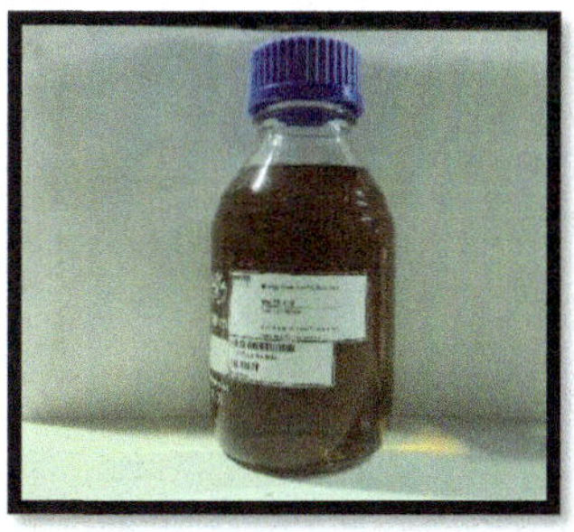

(continued)

Table 3.13 (continued)

Additive name	Abbreviated name	Photo
Surface tension reducer	WA 120 C	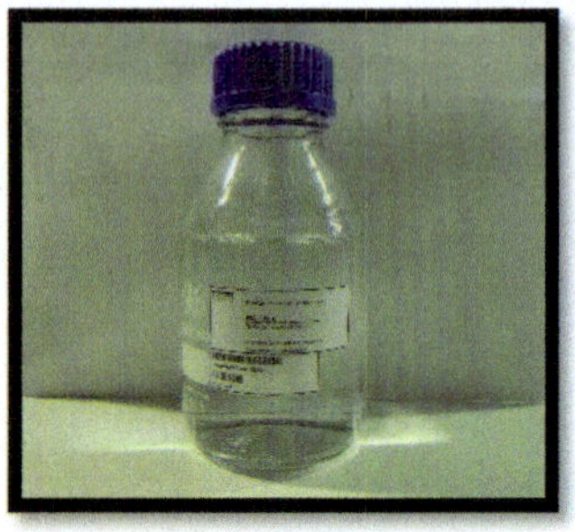
Acid retarder	WA 911	
Corrosion inhibitor	WA 107 plus	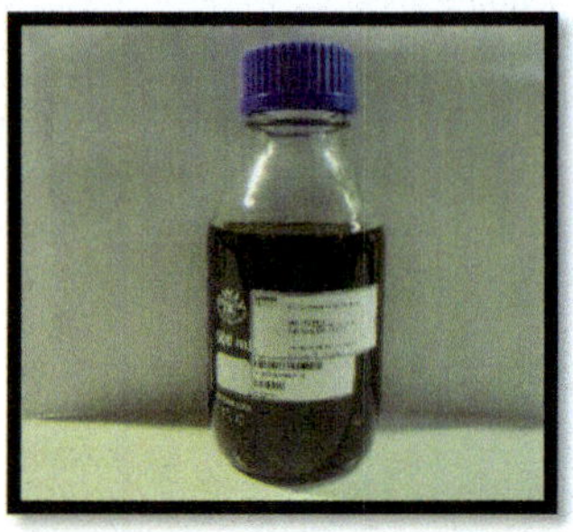

(continued)

Table 3.13 (continued)

Additive name	Abbreviated name	Photo
Acid corrosion inhibitor intensifier aid	WA 810	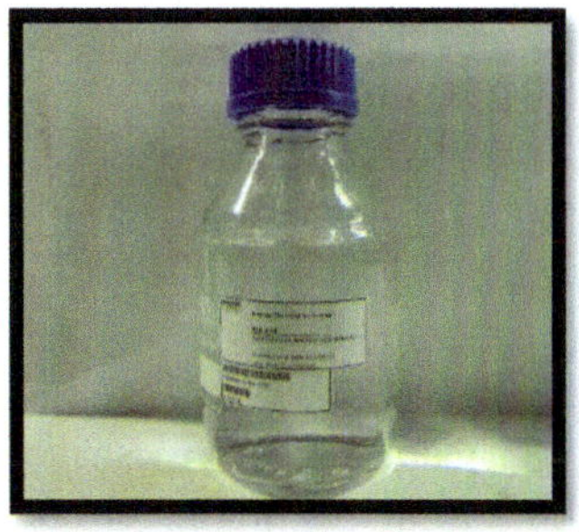
Non-emulsifier	WA 340	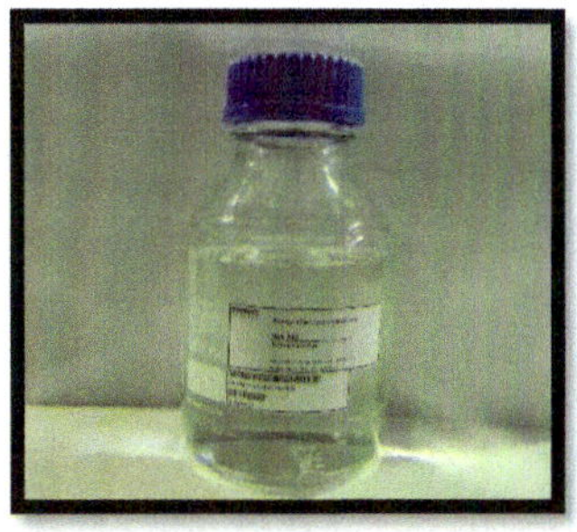
Mutual solvent	WA 212	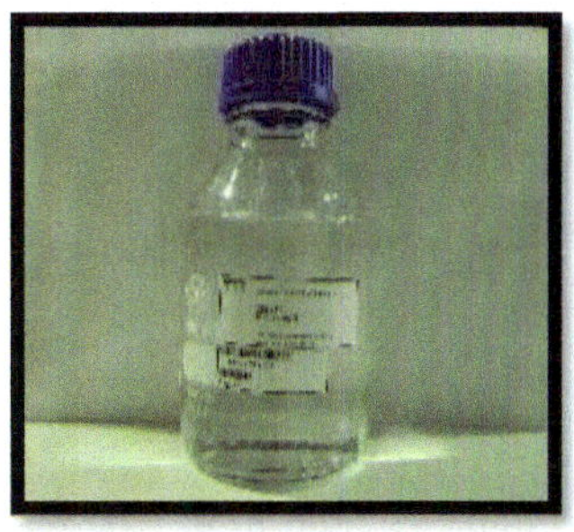

- Iron control methods
  - pH control

pH is controlled by introducing a weak acid with a slow reaction rate to maintain a low pH after the use of HCl. Typically, acetic acid is used for this purpose. A low pH prevents secondary precipitation of iron.

– Separating agents

Separating agents interact with iron to keep it in a soluble state and prevent the formation of deposits. Common release agents include citric acid, acetic diamine ether, and nitrilotriacetic acid.

– Reducing agents

Reducing agents convert ferric (Fe^{3+}) to ferrous (Fe^{2+}). As a result, the pH of the acid does not fall within the pH range conducive to iron precipitation, thus preventing iron-related problems. Common reducing agents include erythorbic acid and sodium erythorbate. Most reducing agents also act as oxygen scavengers, removing dissolved oxygen that would otherwise promote the oxidation of ferrous iron to ferric iron. This helps to keep iron in solution and prevent ferric iron precipitation (Hall & Dill, 1988; Mirvakili et al., 2012; Smith et al., 1969).

- Advantages
  1. Reduces the formation of iron compounds in the reservoir rock.
  2. Maintains permeability of reservoir rock by minimizing deposition in fractures.
  3. Improved acidisation performance by avoiding blockages caused by iron precipitation.
- Disadvantages
  1. Incorrect selection of additive levels may result in inadequate control and a reduction in treatment performance.
  2. Improper treatment with other additives can negatively affect the overall efficiency of the acidizing process.

3.11.2.2 Iron Reducer

Iron reducers are used to solve problems associated with iron compounds in reservoir rock or formation fluids. Iron compounds can precipitate and form solid deposits that block pore spaces within the reservoir. This blockage reduces permeability and impedes fluid flow. Iron reducers work by forming complexes with iron ions, preventing the formation of harmful compounds. By minimizing the risk of formation damage, iron reducers help to improve well productivity.

- Advantages
  1. Reduces formation damage caused by iron compounds, thereby maintaining permeability and fluid flow.
  2. Enhance the effectiveness of acid treatments by preventing pore-clogging due to iron-related problems.

- Disadvantages
  1. Inadequate dosage or improper application may result in incomplete iron reduction, compromising treatment results.
  2. Improper choice of reducing agent may result in the formation of bound iron complexes.
- Application

  Critical for successful reservoir stimulation in formations with iron-related problems.

3.11.2.3 Surface Tension Reducer

Surface tension reducers are additives designed to reduce the surface tension between acid and reservoir fluids. Elevated surface tension can inhibit the ability of the acid to spread evenly throughout the formation. By reducing this tension, the acid can disperse more evenly, ensuring that it reaches a greater surface area of the reservoir rock. This in turn increases rock dissolution and improves well productivity.

- Advantages
  1. Reduces surface tension, facilitating uniform acid distribution throughout the rock formation.
  2. Improves contact with the reservoir, resulting in improved dissolution and increased permeability.
- Disadvantages
  1. Excessive reduction in surface tension can lead to excessive foam production, which can affect fluid flow.
  2. Inappropriate choice of surface tension reducer may result in limited surface tension reduction.

3.11.2.4 Acid Retarders

Acid retarders are chemical additives introduced to regulate the rate of acid reaction with reservoir rock. In certain situations, particularly in high temperature formations, acids can react too quickly, limiting the effectiveness of the treatment. Acid retarders are used to reduce the reaction rate of the acid, allowing it to penetrate deeper into the rock and achieve more complete penetration. This controlled reaction increases the success of the treatment by ensuring a thorough reaction.

- Advantages
  1. Provides a controlled reaction rate, ensuring complete interaction between the acid and the rock.
  2. Allows for longer treatment times, particularly in formations where acid reactions are rapid and beneficial.

- Disadvantages
  1. Excessive retardation of the acid reaction can result in longer treatment times, which may affect operational efficiency.
  2. An inappropriate choice of retarder concentration may disrupt the desired reaction control.
- Application

  Used in high-temperature formations to achieve controlled and effective acidizing.

3.11.2.5 Friction Surface Tension Reducer

Friction reducers are introduced into the acid treatment to reduce friction between the acid and the well surfaces. Friction has the potential to disrupt the flow of acid through the well, limiting its ability to reach the reservoir and interact effectively with the rock. The inclusion of friction reducers facilitates a smoother acid flow, increasing its ability to reach its target efficiently. This in turn contributes to more effective treatment and increased rock dissolution.

- Advantages
  1. Reduces friction between the acid surface and the wellbore, promoting improved acid flow and penetration.
  2. Improves treatment efficiency by facilitating the acid's ability to reach the formation more effectively.
- Disadvantages
  1. Excessive use of friction reducers can lead to unwanted fluid mobility and uncontrolled migration.
  2. Incorrect choice of reducer type may result in inadequate friction reduction.
- Applications

Used to optimize acid flow within the wellbore, ultimately increasing the success of the treatment and the dissolution of the rock.

3.11.2.6 Corrosion Inhibitor

Corrosion inhibitors play a critical role in acidizing operations to protect the integrity of well equipment. When strong acids are injected into the formation, they have the potential to react with metal surfaces and cause corrosion. Corrosion inhibitors in acidizing operations form a protective layer on these surfaces, effectively shielding them from acid-induced corrosion. This maintenance of structural integrity is essential to prevent equipment failure and reduce the need for costly repairs (Ansari et al. 2020a, 2020b).

- Advantages
  1. Protects well equipment from acid-induced corrosion, reducing maintenance and breakdowns.
  2. Maintains the integrity of metal components, ensuring the safe and efficient operation of the well.
- Disadvantages
  1. Insufficient inhibitor dosage can result in equipment damage and an increased risk of corrosion.
  2. Excessive use of inhibitors can potentially interfere with acid-rock interaction and treatment effectiveness.
- Applications

Essential in all acidizing operations to prevent equipment damage and maintain well structural integrity.

3.11.2.7 Non-emulsifiers

Non-emulsifiers play a critical role in acidizing operations by preventing the formation of emulsions between the acid and the crude oil. Emulsions are stable mixtures of two immiscible liquids, such as oil and water that can inhibit the ability of the acid to interact with the reservoir rock. Non-emulsifiers disrupt the formation of these emulsions, allowing the acid to separate from the oil and water phases. By ensuring that the acid remains free to react with the rock, non-emulsifiers contribute significantly to the overall effectiveness of acid treatment.

- Advantages
  1. Prevents the formation of emulsions, ensuring efficient contact between the acid and the rock for optimum formation decomposition.
  2. Enhances the ability of the acid to react with the reservoir and improves permeability.
- Disadvantages
  1. Excessive use of non-emulsifiers can lead to the separation of oil and acid, affecting treatment uniformity.
  2. The selection of an inappropriate non-emulsifier may result in inadequate emulsification prevention.

3.11.2.8 Mutual Solvents

Mutual solvents address the challenge posed by reservoirs containing both oil and water phases. In such scenarios, the immiscibility of these two fluids can hinder the effective penetration of acid into the formation. Mutual solvents act as mediators, dissolving both oil and water to create a more homogeneous mixture. This allows the acid to disperse more effectively throughout the reservoir, enhancing its ability to improve permeability (King & Lee, 1988).

- Advantages
  1. Allows acid to penetrate mixed-phase oil and water reservoirs, promoting improved rock dissolution.
  2. Facilitates better distribution of acid throughout the formation, resulting in more consistent treatment results.
- Disadvantages
  1. Inappropriate choice of mutual solvent can result in limited solubility and reduced enhanced recovery.
  2. Incompatibility with other additives can jeopardize the overall success of the treatment.

3.11.2.9 Alcohols

Alcohols serve several purposes in acidizing operations, including removing water blocks, increasing fluid recovery, retarding acid reactions, and reducing water content in acidizing fluids. The most commonly used alcohols for these purposes are isopropanol and methanol.

- Water blocking removal: Water blocking, which occurs in areas with high capillary forces filling pore spaces, significantly hampers production. The addition of alcohol to acidizing fluids reduces these capillary forces, promoting easier movement of liquid phases and preventing the formation of water blocks.
- Improve fluid recycling: Water and acidizing fluids often have a high surface tension, which limits penetration and recovery of the acidizing solution. Adding alcohol to acid solutions reduces their surface tension, facilitating fluid recycling and permeation.
- Relays acid reactivity: Alcohols have a retarding effect on acid reactivity. The degree of retardation depends on the type and concentration of alcohol added.
- Reduction of water content in acids: Clays have a high water absorption capacity, causing them to absorb water dissolved in acid and swell. Clay swelling reduces the permeability of the formation and hinders production. Alcohol is added to acid solutions to mitigate the effects of this phenomenon. However, to be effective, alcohol must be used at relatively high concentrations, which can increase costs. In addition, the presence of alcohol in an acidic solution is more corrosive than in an alcohol-free acidic solution. In the presence of alcohol, the formation brine can lose its salt content, leading to sedimentation. It's important to test the compatibility of the chosen alcohol with the crude beforehand, typically using methods such as sandpack testing. It's important to note that alcohols will readily react with acids under certain conditions. In the case of organic acids, these reactions are reversible, allowing regeneration of the acidizing fluid. In the case of mineral acids, however, the reaction is irreversible, resulting in a loss of acid.

3.12 Guidelines for Laboratory Evaluation of Acidizing Additives

3.12.1 Compatibility Testing

The main purpose of matrix acidizing in carbonate formations is to improve the production index by removing damage around the wellbore and creating a series of wormholes. Factors such as temperature, pressure, hydrocarbon compounds, formation permeability, acid/formation fluid compatibility, acid/additive compatibility, and inter-additive compatibility should be considered when selecting the correct acid. Compatibility between additives, compatibility of additives with the acid, and compatibility of the injection fluid with the minerals of the formation are some of the things that should be checked before operation.

A number of additives are the basis and constant of any acidizing operation. These include iron controllers, corrosion inhibitors, surfactants, mutual solvents, friction reducers, and breakers. For example, the addition of a corrosion inhibitor is necessary in any acidizing process. This addition prevents corrosion of the mobile brain tube, wall tube, pumps, transmission lines, and agitator reservoirs. Mutual solvents are used in acidizing to facilitate the reaction of acid with hydrophobic rock. The addition of an iron regulator prevents unwanted iron sulfide and hydroxide damage by preventing iron precipitation at low pH. Surfactants are often used as emulsifiers, non-emulsifiers, anti-sludge agents, foaming agents, and water-wetting agents. Non-emulsifiers are often used to break the emulsion that tends to form between hydrochloric acid and hydrocarbons.

Incompatibility between different additives used in acidizing can lead to formation damage. For example, the combination of cationic corrosion inhibitors with anionic surfactants can cause formation problems and must be used with care. In addition, the presence of mutual solvents can reduce the effectiveness of corrosion inhibitors, so if used, the concentration of corrosion inhibitors must be increased. When corrosion inhibitors are added with iron controllers and non-ionic surfactants, they reduce the reactivity of hydrochloric acid with certain minerals.

Friction-reducer additives, being anionic, are incompatible with cationic corrosion inhibitor additives. Certain acidizing operations, particularly in deep wells, require the addition of friction reducers in addition to corrosion inhibitors. Thorough compatibility tests should therefore be carried out prior to use.

The selection of acidizing additives depends on factors such as formation rock mineralogy, formation temperature, permeability, compatibility between additives, and compatibility with reservoir fluids. An acidizing operation may involve all or some of these additives, highlighting the importance of assessing the compatibility of all components to avoid damage and complications.

3.12.2 *Test Method*

Compatibility assessment involves two key steps: firstly, assessing the compatibility of individual additives with the acid, and secondly, assessing the compatibility of the combination of all additives with the acid. To assess the compatibility of each additive with the acid, the amount of acid and water required is calculated. For the compatibility test, 1 ml of the additive is required per 100 ml of solution. This additional volume is subtracted from the volume of water required to make the solution. For example, to make 100 ml of 15% hydrochloric acid with the addition of a corrosion inhibitor, you'll need 45 ml of 33% hydrochloric acid, 54 ml of distilled water, and one milliliter of the corrosion inhibitor. This test is carried out under two temperature conditions: ambient temperature and reservoir temperature, over two time periods of 6 and 24 h. At the end of each time interval, the solution is examined for consistency, solids, sludge, emulsion, sediment, and separation.

3.12.2.1 The Experiment Results

The desired acids are used to check the compatibility test using the results of the solubility test obtained, which are as follows:

- HCl 15%
- HCl 7.5% + HAc 2.5%
- HCl 15% + MeOH
- HCl 7.5% + MeOH

The compatibility of each acid, both individually and in combination with additives, is assessed. As an illustrative example, Fig. 3.57 shows the compatibility of 7.5% HCl + 2.5% HAc acid with a retarder. This figure shows that the retarder does not form a homogeneous mixture but appears as a separate phase on top of the solution, regardless of whether it is at ambient or reservoir temperature.

But Fig. 3.58 for the same acid 7.5% HCl + 2.5% HAc with corrosion inhibitor intensifier aid is homogeneous and has no precipitation and no reaction with added additives.

3.12.2.2 Iron Control Agent

During the acidizing process, the pH of the acid gradually increases. When the pH rises above about 2, Fe^{3+} forms a sticky black deposit which reduces well permeability. The use of a Fe^{3+} reducing or ion-controlling agent as an additive is therefore effective in prevting the formation of this type of deposit. If significant amounts of iron are present in the form of Fe^{3+} rather than Fe^{2+}, acid dissolution will result in iron precipitation and subsequent permeability reduction after acidizing. Fe^{3+} iron precipitates at a pH of about 2, whereas Fe^{2+} iron precipitates at a pH of about

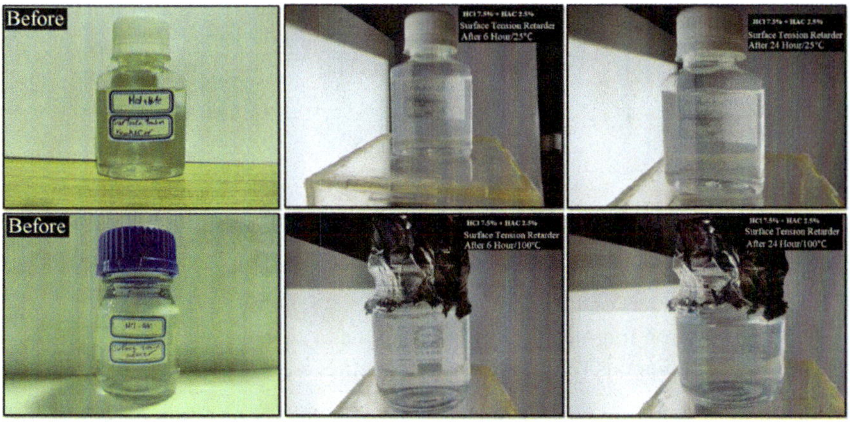

Fig. 3.57 Compatibility of surface tension reducer with 7.5% hydrochloric acid and 2.5% acetic acid

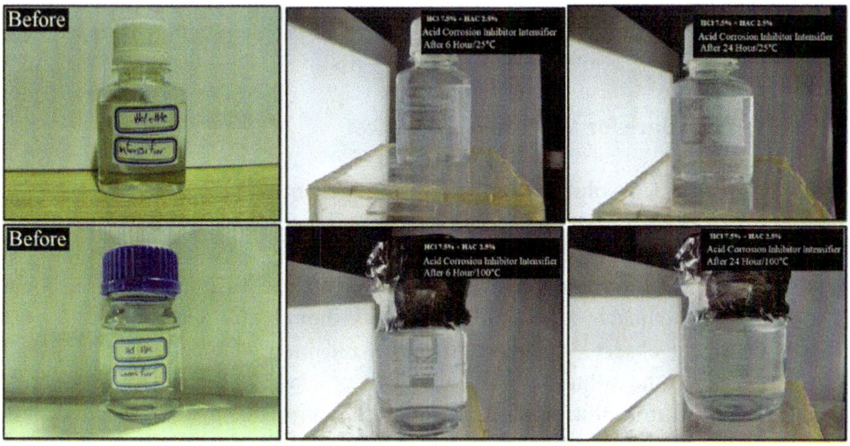

Fig. 3.58 Compatibility of corrosion inhibitor intensifier aid with 7.5% hydrochloric acid and 2.5% acetic acid

7. As spent acid solutions rarely exceed a pH of 6, Fe^{2+} precipitation is rarely a problem. A persistent problem is the presence of iron(II) in the presence of undissolved calcium carbonate. Iron(II) may precipitate slowly as iron(II) carbonate. This doesn't usually impede flow in carbonate rocks but can cause problems in carbonate-rich sandstone formations. To overcome this, the addition of acetic acid as a low pH pre-wash can be beneficial in preventing iron carbonate precipitation in these scenarios. During the acidizing process, one or more of the following iron sources may become problematic:

Table 3.14 Oxidation status of iron in formation minerals

Minerals	Chemical formula	Oxidation state
Hematite	Fe_2O_3	Fe^{2+}
Magnetite	$FeO - Fe_2O_3$	$Fe^{2+} - Fe^{3+}$
Pyrite	FeS	Fe^{2+}
Siderite	$FeCO_3$	Fe^{2+}
Chlorite clay	–	Fe^{2+}

- Corrosion products on the pipe wall
- Oxide crust (iron fragments detached from metal components)
- Iron-bearing minerals (e.g., hematite, magnetite, etc.)

In the Earth's crust, iron is predominantly found in the form of iron oxide minerals, including hematite, magnetite, goethite, and limonite. The primary minerals utilized as ores for iron production are hematite and magnetite. Table 3.14 provides a list of iron-containing minerals found within the formation minerals.

As can be seen from Table 3.14, the iron in these minerals is in the oxidation state Fe^{2+}. Therefore, due to the low pH, there will be no problem with iron deposition.

3.12.2.3 Procedure for the Iron Control Test

First, prepare a 100 ml solution with a specific concentration of ferric ions (500 ppm) using ferric chloride ($FeCl_3$) III and a 15% hydrochloric acid solution. Then add the iron control agent to the acid at the concentration recommended by the manufacturer. Adjust the acidity of the solution to a pH between 1.5 and 2 using calcium carbonate powder and then adjust to a pH of 4 using 1 M sodium hydroxide solution.

If you do not observe the formation of a red precipitate of ferric hydroxide ($Fe(OH)_3$), place the solution in a water bath or oven at 185 °F for 6 h. At the end of the test, if the solution is clear without any precipitate, state the percentage concentration of the said additive. If not, repeat the test with a higher dose of the additive. Determine the optimum dosage of the additive for each required concentration of ferric ion.

Note 1: State the minimum concentration of additive that can effectively maintain 500 ppm iron ions in solution throughout the test period as the optimum dosage of additive.

Note 2: The maximum acceptable concentration of the iron ion regulator is 3% by volume.

Note 3: The presence of white sediment resulting from pH adjustment with calcium carbonate does not necessarily indicate ineffectiveness. However, the formation of sediment of any color (except white) should be investigated.

Note 4: Consider the concentration of iron ions in relation to the final volume of the solution after pH adjustment.

3.12.3 *Iron Reducing Agent*

First prepare 100 ml of a solution containing a specific concentration of ferric iron (500 ppm) using ferric chloride ($FeCl_3$) and a 15% hydrochloric acid solution. To adjust the pH of the solution, add the optimum concentration obtained from the iron ion controller addition test and the concentration suggested by the manufacturer of the iron ion reducer. Adjust the pH of the solution to 4 using calcium carbonate and a molar solution of sodium hydroxide after the iron ion control addition test. Measure the concentration of ferrous ions and total iron using a spectrophotometer. The solution should be stable at 185 °F for 6 h, during which time more than 90% of the ferric iron should be converted to ferrous iron. Report the concentration of this percentage increase as the optimum value for the 500 ppm iron ion level.

Note 1: The maximum acceptable percentage concentration of iron ion reducing agent is 3 vol%.

Note 2: The white precipitate resulting from raising the pH does not necessarily indicate ineffectiveness. However, any sediment of a colour other than white should be analysed.

3.12.4 *Additive Evaluation for Removing Hydrogen Sulfide Gas*

Pour a specified amount of iron sulfide powder (passing through a 100 mesh sieve) into a flask fitted with a side outlet. Connect the outlet of the flask to two acid gas containers placed in series. Add 100 ml of a 10% volume cadmium sulfate absorbent solution to each acid gas container. In a 50 ml capped flask, mix acidic additives, including the hydrogen sulfide remover additive, at a specified concentration. Add 28% hydrochloric acid to achieve the required volume. Place the bottle in place and direct the generated H_2S gas into the sour gas containers using N_2 gas as a carrier. After 2 h, separate the sour gas containers and determine the amount of H_2S by titration of the containers containing cadmium sulfide deposits using the iodometric method. State the amount of H_2S relative to the volume of acid (50 ml).

3.12.4.1 Additive Evaluation

Weigh the iron sulfide powder again to the amount used in the control sample test and repeat the test by adding the H2S remover additive to the other additives in the acid phase. Before introducing the nitrogen gas, keep the outlet of the system to the saline gas containers closed for 1 h to increase the reaction time. Then, by opening the flow of nitrogen gas for 1 h, direct the remaining H_2S gas towards the absorbent containers.

Note 1: In the presence of 500 ppm of H_2S gas, report the lowest concentration of the mentioned additive that can remove at least 90% of the H_2S gas as the optimal concentration of the additive.

Note 2: Remove the maximum acceptable percent concentration of the hydrogen gas remover additive. Hydrogen sulfide gas is 3% by volume.

3.12.5 Evaluation of Suspending Agent

To carry out the test, first, mix a certain quantity of calcium carbonate powder with 15% hydrochloric acid and adjust the pH to 4. Then filter the solution through filter paper. Pour 100 ml of the prepared acid into a graduated pear-shaped container (centrifuge tube), then add 5 g of silica fluor powder and sodium bentonite filtered at 100 mesh in a ratio of approximately 1–4. Mix the contents vigorously for one minute. Then place the container in a fixed position and note the amount of sediment deposited within 30 min (this is the control sample sediment test). Repeat the test with the suspended additive and record the changes in sediment volume over time for both tests. As an additional assessment, measure turbidity using a spectrophotometer over 30 min and compare the turbidity results between the two control samples and the sample containing the suspended preservative.

Note 1: Report the highest concentration of additive capable of suspending 75% of the sediment from the control sample as the optimum concentration.

Note 2: To prepare the acid used, add sufficient calcium carbonate powder to 15% hydrochloric acid to maintain a pH of about 4.

Note 3: The maximum acceptable percentage concentration of the anti-suspension additive is 3% by volume.

3.12.6 Anti Sludge Evaluation

Take 50 ml of hydrochloric acid (15 or 28%), add the anti-sludge additive and other specific additives at the recommended doses, and introduce 500 ppm Fe^{3+} ions using FeCl (with the addition of salt) in a 100 ml cylinder. Then add 50 ml of crude oil from the indicator well. Mix the contents for 30 s using a blender at 14,000 rpm (Hamilton Beach). Place the bottle in an oven at 185 °F. After 4 h, pour the contents of the cylinder onto a 100-mesh screen. If you see sludge on the mesh, rinse it with diesel and hot water. If sludge remains, repeat the test with a higher dose of anti-sludge additive (standard reference RP 42).

Note 1: Ensure that other acid additives (corrosion inhibitors, corrosion inhibitor boosters, iron control agents, iron ion reducers, non-emulsifiers, suspending agents, and H_2S removers) are used at their optimum percentages.

Note 2: The lowest additive concentration capable of removing sludge at the end of the test period should be reported as the optimum additive concentration.

Note 3: The maximum acceptable concentration of anti-sludge additive is 3 vol%.

3.12.7 Evaluation of Non-emulsifiers

Pour 2.5 g of powder (90% silica fluor + 10% bentonite after passing through 100 mesh) into a container and add 50 ml of index well oil and 50 ml of hydrochloric acid (15 or 28%) along with non-emulsifier in the presence of Add other additives at the specified dose and 500 ppm of Fe ions (using salt) $FeCl_3$ to the container and mix for 30 s with a stirrer at a speed of 14,000 rpm. Transfer the contents of the container to a 100 ml cylinder and place in the oven at 185 °F.

Record and report the amount of phase separation at 5-min intervals for 30 min. Repeat the test with the standard reference (RP 42) and report the results.

Note 1: To spend 15% hydrochloric acid, add enough calcium carbonate powder to prevent the acid from boiling and adjust the pH to 4.

Note 2: State the minimum concentration of the non-emulsifying additive capable of completely separating the two phases within 30 min as the optimum concentration of the above additive.

Note 3: The maximum acceptable concentration of non-emulsifiers is 3% by volume.

3.12.8 Acid Corrosion Inhibitor and Intensifier Evaluation

One of the most important additives in acid treatment is corrosion inhibitors. Corrosion inhibitors are chemicals designed to reduce the rate of acid-induced corrosion on drill pipes and wellhead tubing. The simplest and most reliable method of estimating corrosion is by weight loss analysis. In this method, a sample (called a coupon) is immersed in the solution for 12 h, then removed, cleaned to remove any signs of corrosion, and reweighed. The corrosion rate (CR) is calculated by the weight loss method according to Eq. 3.12.

Corrosion Rate (CR):

$$CR = \frac{\text{weightloss(gr)} * \text{K}}{\text{AlloyDensity}\left(\frac{\text{g}}{\text{cm}^3}\right) * \text{ExposedArea}\left(cm^2\right) * \text{ExposureTime(hr)}} \tag{3.12}$$

Table 3.15 K factor in different units for corrosion rate

Desired corrosion rate unit	Surface unit	Factor—K
One-thousandth of an inch per year (mpy)	in^2	5.34×10^5
One-thousandth of an inch per year (mpy)	cm^2	3.54×10^6
mm per year (mmy)	cm^2	8.76×10^4

Table 3.15 shows the K constant for different units to calculate the corrosion rate.

3.12.8.1 Test Method

Prepare the required volume of acid solution with an approximate ratio of 5.6 ml/cm^2 of acid volume to coupon surface area. Add the corrosion inhibitor, corrosion inhibitor intensifier, and any other required acid additives in specified doses. Transfer this mixture to the container of the high temperature and high pressure (HPHT) corrosion rate measuring apparatus. Adjust the temperature and pressure settings of the apparatus as required (usually 1000 psi pressure and the desired temperature).

The metal coupons used in this test are typically made from well wall tubing, such as N-80. Measure the initial weight and surface area of all sides of the coupons. Suspend these coupons in the acid solution with the additives and then place the container containing the coupons and acid in the machine. After 6 h at the required temperature and pressure, remove from the machine.

Wash, dry, and weigh the coupons. Finally, calculate the corrosion rate in Ib/ft^2 (the test is carried out in static mode). Examine the coupon surface with a magnifying glass and report any holes or irregular corrosion on the appropriate form.

Repeat the tests to check the corrosion rate and to determine the optimum doses of corrosion inhibitor and corrosion inhibitor intensifier in the presence of other additives for hydrochloric acid at concentrations of 15 and 28% and other desired acids. Report the optimum dose of corrosion inhibitor and corrosion inhibitor intensifier.

If necessary, the corrosion inhibitor test and the determination of the optimum dose of corrosion inhibitor and corrosion inhibitor intensifier under dynamic conditions will be carried out at a later stage, after appropriate preparations have been made for this test.

Note 1: For corrosion inhibitor additives and corrosion inhibitor intensifier aids, report the minimum concentration that results in a corrosion rate of less than 0.05 lb/ft2 as the optimum dose for these additives.

Note 2: For other acidic additives (iron ion controllers, iron ion reducers, anti-sludge agents, anti-emulsifiers, suspending agents, and H_2S effect removers), use the optimum percentage concentration of each additive. In addition, for additives dependent on oil type, determine the optimum dose in the presence of well oil samples as an indicator.

Note 3: If unusual pitting is observed on the coupon, even if the corrosion rate is less than 0.05 lb/ft^2, the test should be repeated with a higher dose of corrosion inhibitor to eliminate the pitting.

Note 4: A corrosion inhibitor intensifier is not required for tests to determine corrosion rate at temperatures of 200 °F and below.

Note 5: When evaluating with 15% hydrochloric acid, first add the optimum percentage concentration of each additive to a specified volume of distilled water, stir well, then add concentrated hydrochloric acid to achieve a final acid concentration of 15%.

Note 6: After determining the optimum concentration of the corrosion inhibitor, keep the concentration of the corrosion inhibitor constant and then optimize the percentage concentration of the corrosion inhibitor intensifier.

Note 7: The maximum acceptable percentage concentration for the corrosion inhibitor is 3% by volume, and for the corrosion inhibitor intensifier it is 4% by volume.

3.12.8.2 Test Results

Figures 3.59, 3.60, 3.61, and 3.62 show the corrosion test results for hydrochloric acid at a concentration of 1.5% corrosion inhibitor intensifier and corrosion inhibitor. Table 3.16 shows the corrosion test results for four acids.

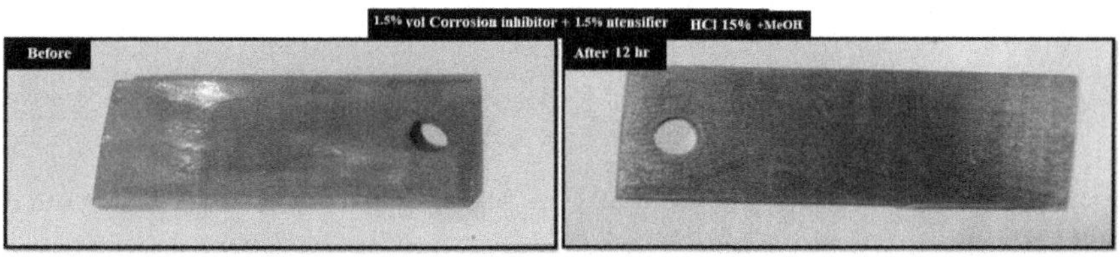

Fig. 3.59 Corrosion of N-80 coupon after 12 h at reservoir temperature by 15% hydrochloric acid and methanol

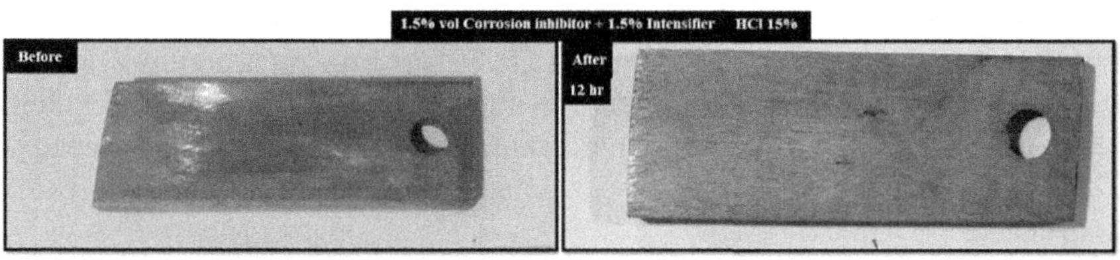

Fig. 3.60 Corrosion of L-80 coupon after 12 h at reservoir temperature in 15% hydrochloric acid

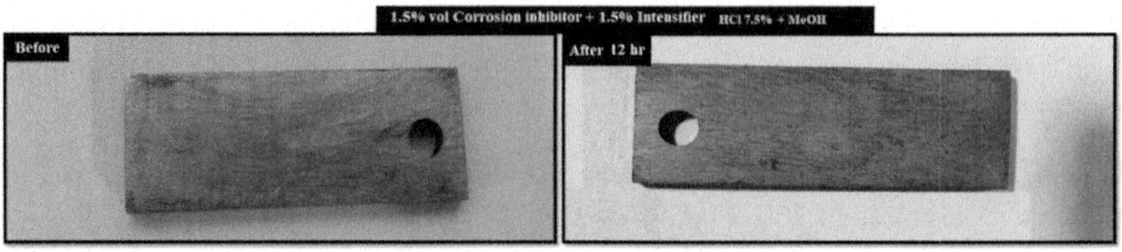

Fig. 3.61 Corrosion of L-80 coupon after 12 h at reservoir temperature by 7.5% hydrochloric acid and methanol

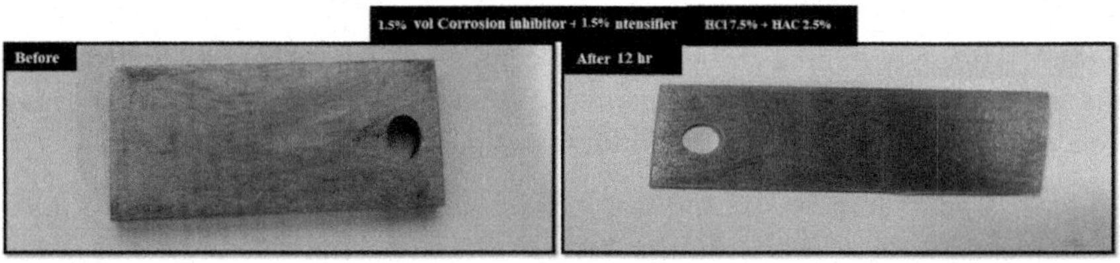

Fig. 3.62 Corrosion of L-80 coupon after 12 h at reservoir temperature by 7.5% hydrochloric acid and 2.5% acetic acid

Table 3.16 Corrosion test results

Sample	Corrosion inhibitor concentration (volume percentage)	Corrosion inhibitor intensifier aid concentration (volume percentage)	Time (h)	Corrosion rate (mpy)
HCl 15%	1.5	1.5	12	103.8752
HCl 7.5% + HAc 2.5%	1.5	1.5	12	116.9682
HCl 7.5% + MeOH	1.5	1.5	12	163.1
HCl 15% + MeOH	1.5	1.5	12	272.3292

Based on the test results and examination of the coupon surfaces, it is evident that there was no corrosion on these surfaces. Furthermore, in accordance with the standard, the corrosion rate is less than 500 mpy (mils per year). The concentration selected to increase the corrosion inhibitor remains at 1.5%. In addition, the comparison of corrosion in these four acids shows an increase in corrosion in the presence of methanol.

3.12.9 *Review of Surface Tension Reducer*

In reservoirs, the presence of immiscible fluids gives rise to a phenomenon known as surface tension. This surface tension represents the force between immiscible phases. The lower the surface tension, the easier it is for the fluids to move and even become miscible. In acidizing operations, the surface tension between the acid and the hydrocarbon acts as a barrier to the acid penetrating deeper into the rock. By introducing a surface tension reducer, the surface tension between these two phases can be reduced.

3.12.9.1 Test Method

Prepare solutions containing the desired volume percentages of the above additives in distilled water, e.g. 0.5, 1, and 1.5%. Measure the density of these prepared solutions and of the indicator oil. Fill the bulk liquid reservoir of the IFT with one of the prepared solutions and the drop reservoir of the instrument with indicator oil. Set the temperature and pressure of the instrument to the desired values, specifically a temperature of 200 °F and a pressure of 2000 psi.

When the instrument has reached the set temperature, use the pump and valve in the drop reservoir to slowly create a suspended oil drop in the solution. Then enter the densities of the crude oil and the additive solution into the software. Record the interfacial tension (IFT) of the two fluids when equilibrium is reached and maintained.

In the surface tension test, different concentrations of the additive are mixed with the acidizing fluid. The four desired fluids are prepared according to the specified volume percentages. The test is initially carried out at ambient pressure and temperature. By lowering the surface tension, which reduces the tension between the two phases, the movement of the two fluids is facilitated. The hanging drop method is chosen to measure surface tension because of its common use and high accuracy. In this method, a drop of the mixture is placed in a medium containing the desired acid. The properties of the drop are recorded in a photograph and the surface tension is determined by calculating the parameters outlined in Eq. 3.13 (Fig. 3.63).

$$\sigma = \frac{\Delta \rho g {d_e}^2}{H} \quad (3.13)$$

σ: Surface tension, $\frac{\text{dyne}}{\text{cm}}$

$\Delta\rho$: Density difference between injection fluid and ambient, $\frac{\text{g}}{\text{cm}^3}$

g: Gravitational constant, $\frac{\text{cm}}{\text{s}^2}$

d_e: Minimum horizontal diameter of unmagnified droplet, cm.

d_s: The measured diameter of the droplet at a distance d_e from the top of the droplet.

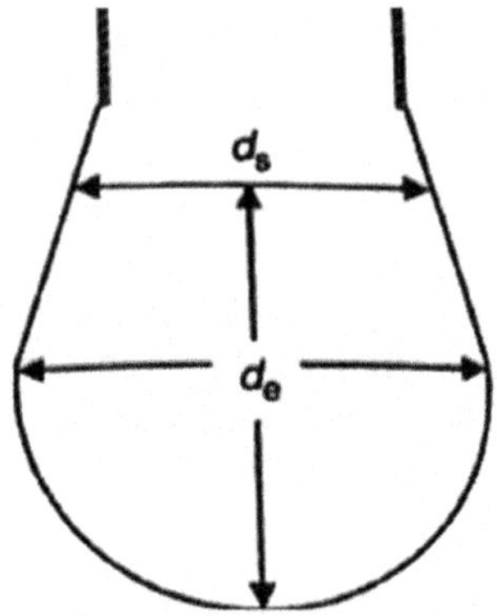

Fig. 3.63 The shape of a hanging drop (Berry et al., 2015)

H: Drop shape coefficient as a function of $\frac{d_s}{d_e}$

If, after calculating the surface tension, the concentration is less than 4 dyne/cm, this concentration can be selected as the optimum concentration.

3.12.9.2 Test Results

The surface tension test was carried out on four acids: HCl 15%, a mixture of HCl 7.5% and HAc 2.5%, HCl 15% with MeOH and HCl 7.5% with MeOH. In this test the surface tension was measured at four different percentages by volume of surface tension reducer: 0.5, 1, 1.5, and 2%. Figure 3.64 shows an example of this surface tension test carried out at 25 °C between condensate, hydrochloric acid, and methanol with increasing levels of surfactant.

3.12.10 Experiment to Determine the Optimal Concentration of Mutual Solvent Additive in the Pre-injection Fluid

When acidizing carbonate rocks, the effectiveness of the acid-rock reaction can be significantly reduced if the reservoir rock has hydrophobic properties. In such cases, the hydrocarbons present on the rock surface can act as a barrier, preventing the acid, which is an aqueous solution, from coming into direct contact with the rock. This reduced interaction between the rock and the acid leads to a reduction in the efficiency of the acid.

To increase the reactivity of the acid with the minerals in the rock, it is necessary to alter the wettability of the rock to make it more amenable to acid-rock reactions. One way to achieve this is to change the wettability from hydrophobic to hydrophilic by the addition of mutual solvents. This change in wettability prepares the rock surface for effective acid-rock reactions. When mutual solvents are introduced with

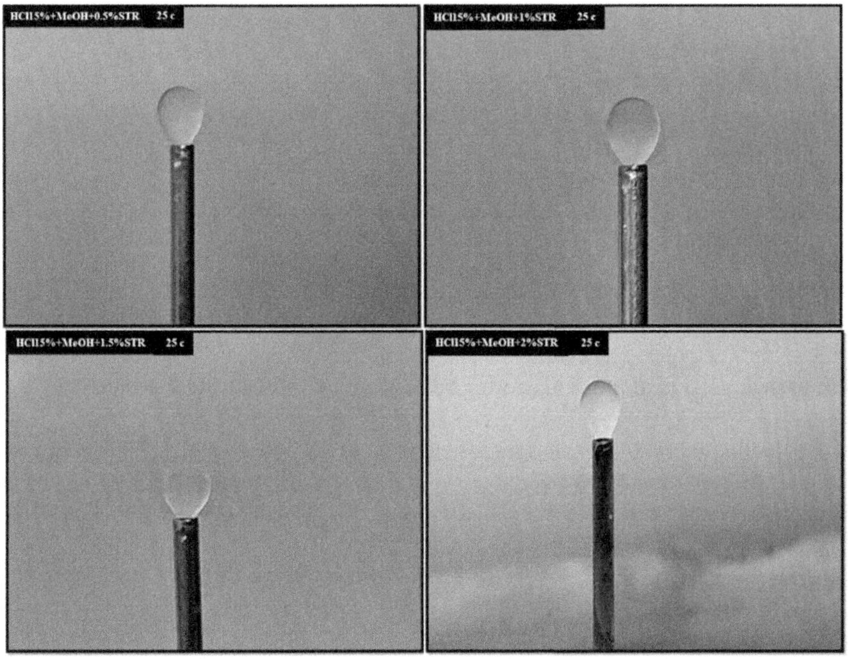

Fig. 3.64 Surface tension between condensate, hydrochloric acid and methanol with increasing surface tension reducer at 25 °C

the pre-injection fluid, they enhance the efficiency of the primary injection fluid, thus optimizing the rock reservoir for acid treatment.

Mutual solvents, such as ethylene glycol monobutyl ether, are selected for their ability to dissolve in both aqueous (acid) and hydrocarbon (oil) fluids. They are used to reduce the surface tension of the reactive acid and speed up the recovery of the used acid during the cleaning of the well. It's worth noting that the use of reciprocal solvents can affect the effectiveness of corrosion inhibitors, often requiring an increase in inhibitor concentration during acidizing treatments when reciprocal solvents are used.

The successful use of crosslinking solvents can vary and their addition typically involves concentrations of around 10% by volume of the acid. This can therefore contribute significantly to the overall cost of the recovery process. Some of the main applications of cross dissolvers include:

1. Water saturation reduction: Mutual solvents help to reduce water saturation around the well by lowering the surface tension of the water, thus preventing water blockage.
2. Formation Wettability Enhancement: They help to create a hydrophilic structure within the formation, maintaining optimum relative permeability.
3. Inhibitor and Insoluble Dissolution: Mutual solvents facilitate the dissolution of inhibitors and insoluble materials in the acid. Some acid corrosion

inhibitors contain acid-insoluble materials that can block the formation and cause adsorption of the inhibitor to formation minerals, altering wettability.

4. Concentration Maintainers: They help maintain the required concentration of surfactants and inhibitors in the solution by reducing the absorption of these substances.
5. Emulsification: Solvents can act as emulsifiers.
6. Improved acid cleaning: They help to improve the cleaning of used acid after the acidizing process.

The use of mutual solvents, both as a pre-wash and in conjunction with acid, is widely accepted for its effectiveness in preventing and breaking emulsions, sludges, and water blockages. The central idea behind mutual solvents is their ability to reduce the surface tension of the acid, from 72 dynes/cm to less than 40 dynes/cm and even below 10 dynes/cm.

Typically, mutual solvents are used in acidizing processes at concentrations ranging from 10 to 35% by volume. Figure 3.64 illustrates the changes in surface tension observed with different concentrations of mutual solvent in combination with hydrochloric acid (Fig. 3.65).

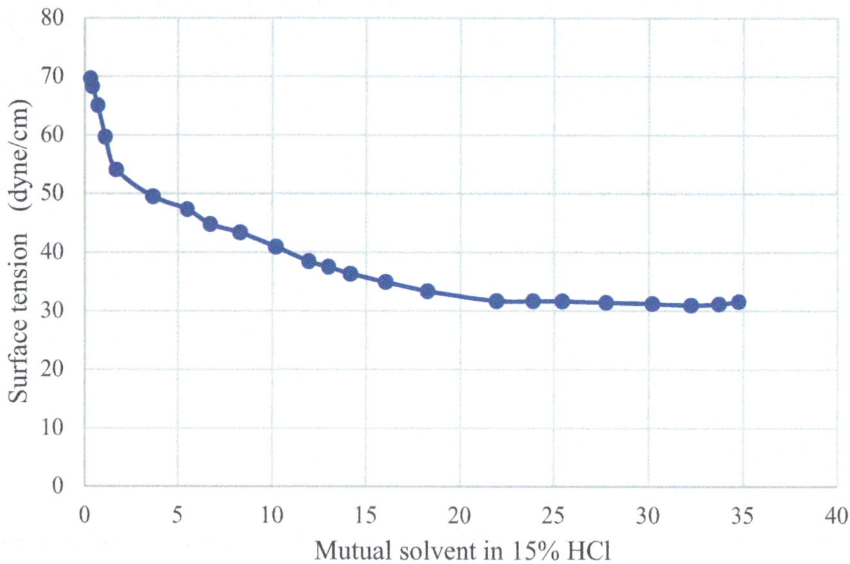

Fig. 3.65 Changes in the surface tension of hydrochloric acid in the presence of different percentages of mutual solvent

3.12.10.1 Solubility Check

First, the solubility of the desired additive at a 10% volume concentration is evaluated in two fluids: water and diesel. If the additive dissolves completely in both aqueous and organic base fluids, the contact angle measurement test is carried out on carbonate rock in the presence of a mutual solvent.

3.12.10.2 Test Method

The test method for determining the optimum concentration of a mutual solvent begins by saturating a thin section cut from the plug with the hydrocarbon fluid from the formation. This saturation process takes 48 h at the temperature of the formation. At the end of this period, the thin section is placed in a contact angle measuring device where the contact angle is measured by injecting a drop of distilled water into an environment containing the mutual solvent, condensate/diesel/acid.

This procedure is repeated for different volume concentrations of the mutual solvent, with the maximum volume percentage set at 10%. The contact angle is calculated by taking a photograph within 2 h. The minimum concentration giving a contact angle of less than 60° is chosen as the optimum concentration.

3.12.10.3 Test Results

To determine the optimum concentration of the mutual solvent, it is essential to measure the contact angle for each concentration. Contact angles were measured for five different states: zero percent, one percent, one and a half percent, two percent, and four percent by volume. A drop was placed on the rock for two hours in the presence of both diesel and the mutual solvent. The contact angle of the drop was recorded every half hour. When the contact angle for a particular concentration fell below 60°, that concentration was considered to be the optimum choice. Figures 3.66, 3.67, 3.68, 3.69, and 3.70 show some examples of the contact angle tests.

3.12.11 Emulsifier Agent Acid Additive Evaluation

To evaluate the effect of increasing the concentration of the emulsifier, 200 ml of emulsifying acid is prepared with a mixture containing 70% by volume hydrochloric acid, 15% by volume distilled water, and 15% by volume diesel fuel. To prepare 140 ml of 15% concentrated hydrochloric acid, the required amount of distilled water is calculated and added to a cylinder, followed by the addition of other required additives (excluding the emulsifier). After dissolving these additives in water, the volume is adjusted to 140 ml with concentrated hydrochloric acid. In a separate container, the desired percentage concentration of emulsifier is poured in relation

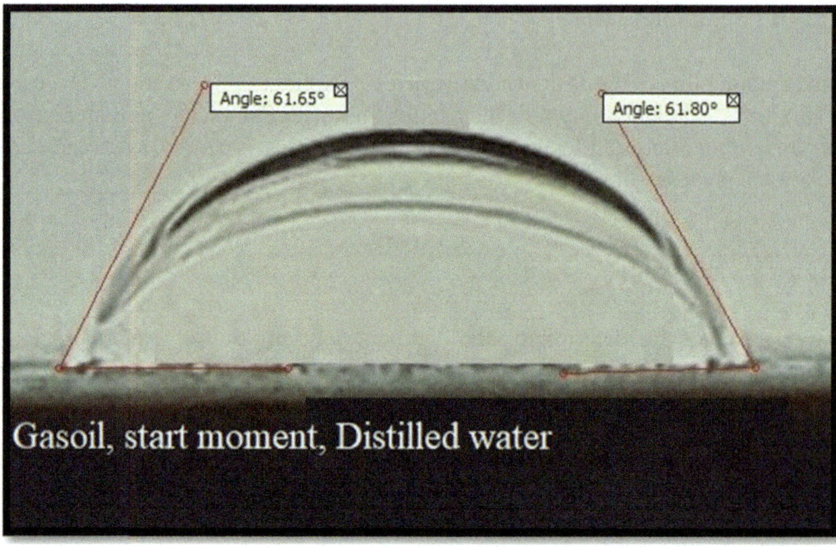

Fig. 3.66 Contact angle between diesel and distilled water without mutual solvent at initial time

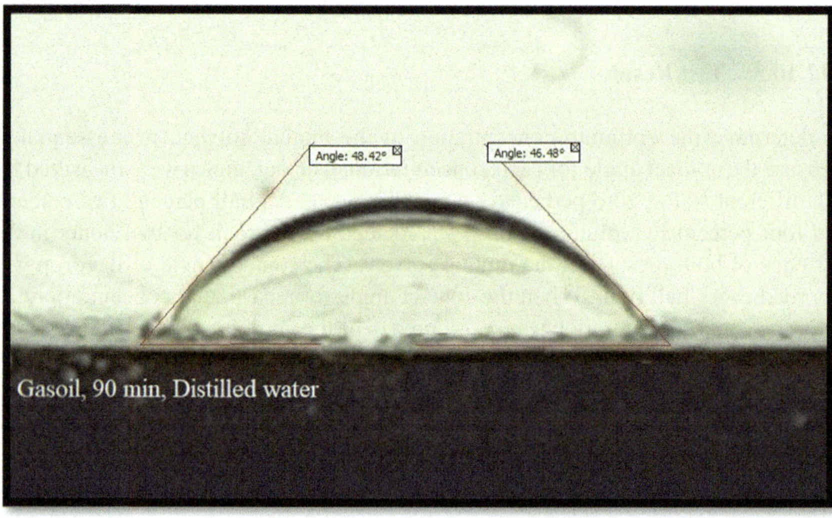

Fig. 3.67 Contact angle between diesel and distilled water without mutual solvent in 90 min

to the total volume of hydrochloric acid and diesel fuel, and diesel fuel is added to make a total volume of 60 ml. Shake well and transfer to a beaker. Using a magnetic load and a stirrer set at 500 rpm, the acid solution and its additives are slowly added to the mixture over 150 s, with a final stirring period of 30 s. The prepared emulsion

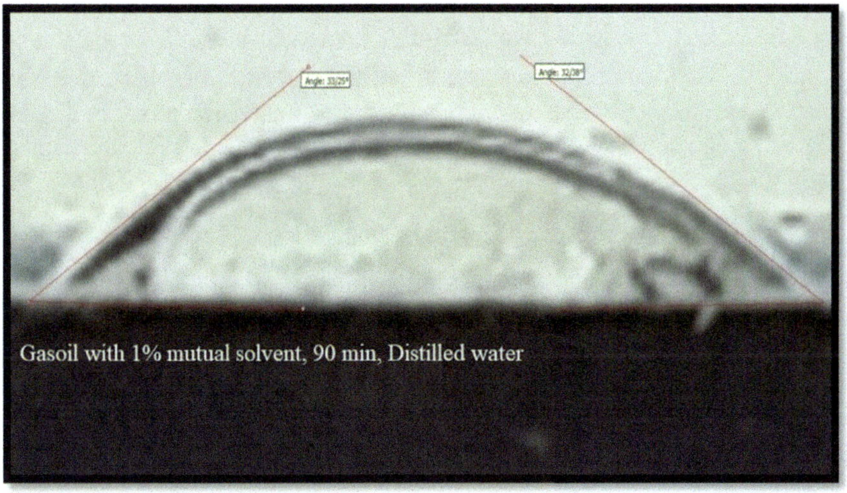

Fig. 3.68 The contact angle of distilled water in the presence of diesel and 1% mutual solvent after 90 min

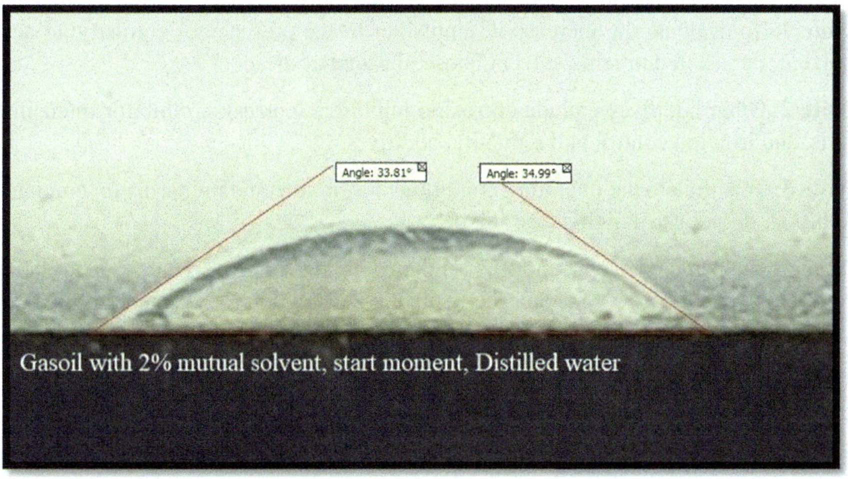

Fig. 3.69 The contact angle of distilled water in the presence of diesel and 2% mutual solvent at the initial moment

acid is then transferred to a 250 ml graduated cylinder which is placed in a water bath at a temperature of 185 °F.

The prepared emulsion acid is monitored continuously for 4 h, recording any observed separation and the number of phases at 30-min intervals. In addition, the electrical conductivity of the emulsifying acid is measured with an electrical conductivity meter; if the electrical conductivity is zero, this indicates an acceptable emulsion

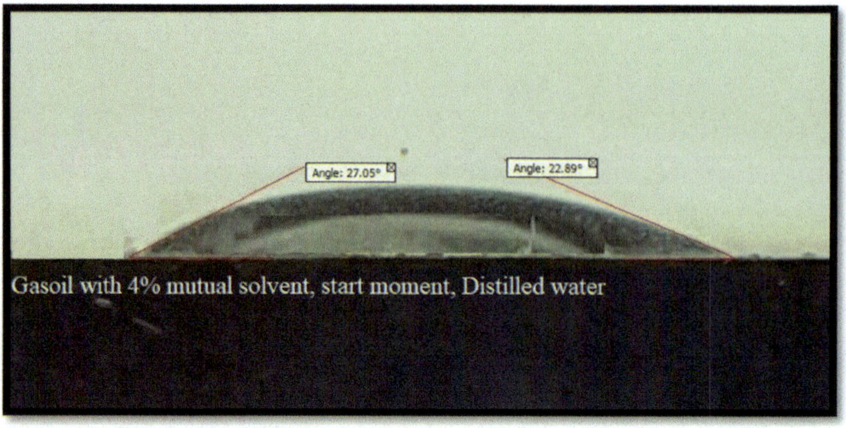

Fig. 3.70 The contact angle of distilled water in the presence of diesel and 4% mutual solvent at the initial moment

quality. The test is repeated at room temperature and the stability of the emulsion within 4 h is assessed and reported.

Note 1: To evaluate the increase in emulsifier in the presence of emulsifying acid solvent, the desired organic solvent is used instead of diesel.

Note 2: Other additives include corrosion inhibitors, corrosion inhibitor intensifier aids, and iron ion control and reduction additives.

Note 3: To simulate the operating conditions for mixing and preparing an emulsion solution, do not use a high-speed mixer.

Note 4: The percentage concentration of the emulsifier additive is calculated relative to the total volume (200 ml) and the percentage concentration of other additives is calculated relative to the volume of hydrochloric acid (140 ml).

Note 5: The minimum percentage concentration of the emulsifying additive, using diesel and other solvents, capable of producing a fully stable emulsion for up to 4 and 2 h respectively, shall be reported as the optimum dosage.

Note 6: The maximum acceptable percentage concentration of emulsifier is 3% by volume.

3.12.12 Evaluation of Clay Stabilizer

3.12.12.1 Swelling Test

This evaluation method follows the ASTM D5890 standard and consists of the following steps:

1. Grinding the bentonite with a mortar and passing it through a 100 mesh screen.
2. Place the ground powder in an oven at a temperature of 105 ± 5 °C for 24 h until completely dry.
3. Pour a specific concentration of clay stabilizer into two 100 ml containers, then add distilled water and 15% hydrochloric acid to make up the required volume. Shake vigorously for 30 s.
4. Control samples in two separate 100 ml bottles by adding 90 ml of distilled water and 15 vol% hydrochloric acid.
5. In two further separate cylinders, add 90 ml of a water solution and 15% hydrochloric acid by volume, together with the additive mentioned in the previous step (as sample solution).
6. Weigh 2 g of dried bentonite powder and divide it into four equal parts.
7. Take each 0.1 g portion of bentonite and gently pour it onto the surface of the liquid in the cylinders and the sample solutions.
8. Allow each 0.1 g to drain for at least 30 s to prevent accumulation of particles and clumping. Allow the bentonite powder to soak and settle slowly for at least 10 min before adding the next portion (in grams).
9. After adding the last portion of bentonite, rinse the particles adhering to the cylinder walls with the remaining liquid until the solution reaches a volume of 100 mm.
10. Place the cylinders in a stationary position without any movement and record the height of the sediment at intervals of 0, 15, 30 min and 1, 2, and 3 h.
11. Calculate the degree of swelling for the bentonite in all four cylinders by dividing the volume of sediment by the total volume of liquid at different times.
12. Record the results obtained. The swelling values obtained for the samples with the addition of clay stabilizer should be lower than those of the control samples.

Note 1: API bentonite powder should be used for evaluations.

Note 2: The minimum concentration of the mentioned additive that prevents swelling should be reported as the optimum dosage of the additive.

3.12.12.2 Lump Stability Test

The clay stability test evaluates the stability of clay lumps under ambient temperature and pressure conditions in both control and clay-stabilized samples. The evaluation is based on the following steps:

1. Select four clay lumps with the same dimensions and contact area.
2. Prepare four separate cylinders, each containing 100 ml of distilled water and 15 vol% hydrochloric acid as control samples. In addition, prepare cylinders containing the same liquids but with a specific concentration of clay stabilizer as sample solutions.
3. Place the clay lumps in the containers with the control samples and the solution containing the clay stabilizer in the designated samples.
4. Record the time from when the clay pellets are immersed in the solution, noting both the beginning and the end of the degradation process.
5. For all four samples, report the type of degradation, describe the structure of the lumps and rate their stability as stable, relatively stable or unstable.

Note: When evaluating this additive, if there is a discrepancy between the optimum interval between the swelling control test and the lump stability test, or if the required additive cannot perform both tests, take the swelling control test as the reference and consider its consumption concentration as the optimum clay stabilizer concentration.

3.12.12.3 Evaluation of Rheological Properties

Measure out 17.5 g of bentonite and combine it with 350 ml of distilled water. Thoroughly mix the solution using a mixer running at a speed of 18,000 rpm for 15 min. Immediately following this, transfer the solution to a rotary viscometer. Switch on the device and measure the viscosity of the solution at two speeds: 600 and 300 rotations per minute (rpm). After turning off the device and allowing it to sit for 10 s, switch it back on and measure the viscosity at 3 rpm. Calculate the values for apparent viscosity, plastic viscosity, and yield point using the following formulas. (R_{600}, R_{300}) represent the speeds of 600 and 300 rotations per minute, respectively.

$$\text{Apparent Viscosity}\quad AV = \frac{R_{600}}{2}$$

$$\text{Plastic viscosity}\quad PV = R_{600} - \partial_{300}$$

$$\text{Yield point}\quad Y_p = R_{300} - PV$$

Repeat all the evaluation steps in the presence of increasing amounts of clay stabilizer at the desired concentrations in distilled water and compare the results of the rheological properties obtained with those of the control sample.

3.12.12.4 Fluid Loss Test

Transfer the solutions prepared in the rheological laboratory to the filter press using suitable filter paper. Seal the chamber lid and apply a gas pressure of 100 psi for

30 min. Place a graduated cylinder under the outlet of the apparatus and record the volume of fluid filtered at the end of the 30 min. Then inspect the apparatus and record the thickness and quality of the mud remaining on the filter paper for both the control sample and the sample containing the clay stabilizer.

Note 1: The more effective the clay stabilizer, the greater the volume of output fluid compared to the control solution and the drier the sludge remaining on the filter paper.

Note 2: The highest consumption concentration obtained from the swelling (sedimentation), lump stability, rheological properties, and loss tests should be reported as the optimum dose of clay stabilizer. If there is a significant difference between the results, the results of each test should be reported separately.

Note 3: The maximum acceptable concentration of clay stabilizer is 3% by volume.

3.12.13 Viscoelastic Evaluation of Surfactant

3.12.13.1 Investigation of Rheological Properties in Spent Acid

1. First, add the required additives to a predetermined amount of distilled water. Next, add the calculated amount of impure hydrochloric acid to the water-additive mixture to make 100 ml of 15% hydrochloric acid and mix thoroughly.
2. Gradually add the viscoelastic surfactant to the solution and mix.
3. Adjust the pH of the prepared acid to 4 either by using carbonate or by adding calcium carbonate powder, effectively preparing spent acid.
4. Under ambient conditions, record the viscosity property at a shear rate of 100 s^{-1}. The optimum dosage of VES should be noted when the viscosity value reaches at least 1000 cp and this property remains stable for at least 10 min.
5. Measure the viscosity of the provided fluid using a rheometer at temperatures of 200 and 275 °F at a speed of 60 rpm. Allow 40 min to reach the desired temperature.

3.12.13.2 Elasticity Measurement

Start by placing a portion of the sample onto the rheometer plate. Adjust the device's temperature to both 200 and 275 °F. Next, identify the region of linear elastic viscosity by applying incremental strains at a frequency of 100 radians per second. This determination is made by analyzing the elastic modulus about strain. Once the linear viscoelastic region is defined, select a strain value within this region to conduct the frequency reversal test. After preparing the device, proceed with the frequency back-and-forth test. Maintain a constant strain within the linear region and ensure the sample is at the desired temperature. In the final step, measure the elastic and viscous moduli with respect to frequency. Evaluate the material for positive elasticity behavior. It's possible that the material's elastic behavior significantly outweighs its viscous behavior ($J' > J''$).

3.12.13.3 Investigation of VES Gel Failure in Different Fluids

Slowly add viscoelastic surfactant (VES) to a solution equivalent to 15% hydrochloric acid by volume, in the presence of other necessary additives, to prepare the desired gel. In six separate graduated cylinders, mix the prepared gel in a 50:50 ratio with different fluids (15 and 28% hydrochloric acid, crude oil, salt water saturated with calcium chloride) and the optimum amount of each solvent and diesel. Measure and report the viscosity of these solutions at 100 rpm.

Note 1: The evaluation of VES must be made in the presence of a VES corrosion inhibitor, VES corrosion inhibitor intensifier (high temperature only), VES iron ion controller, and VES hydrogen sulfide inhibitor.

Note 2: The acceptable consumption concentration for other additives used with VES should be determined separately according to the test method specified for each additive. The maximum acceptable percentage concentration for these additives is 3 vol%.

Note 3: To prepare a solution equivalent to 15% hydrochloric acid, use distilled water together with calcium chloride salt to ensure that the concentration of calcium and chlorine ions is the same as in the acid.

Note 4: The maximum acceptable percentage increase in concentration for VES is 10% by volume.

3.12.14 Foaming Agent Evaluation

To evaluate the foaming agent, begin by adding a specific concentration of the foaming agent additive to a container. Dilute this mixture to a total volume of 100 ml with distilled water and use a stirrer to ensure thorough distribution of the additive.

Next, take a dedicated HPHT glass cell and introduce the appropriate volumes of liquid and gas phases according to the desired foam quality. With the gas and liquid phases combined in the container, set the stirrer to 2000 rpm and mix the contents for one minute. Immediately after stopping mixing, record the initial height of the foam. Then measure the half-life time of the foam, i.e. the time taken for the height of the foam to be reduced by half. If possible, consider producing a color illustrated report to effectively document the results of this foaming agent evaluation.

Note 1: The foam quality percentage is determined by comparing the gas volume to the total volume of the liquid and gas phases.

Note 2: The half-life time is calculated from the time taken for the foam height to be reduced by half.

Note 3: State the minimum percentage concentration of the named additive required to produce a stable foam of the desired quality (70%) in the presence of water, ensuring that the foam half-life is 10 min, which is the optimum additive concentration.

Note 4: The maximum acceptable percentage concentration of the foaming additive is 3% by volume.

3.12.15 Hydrochloric Acid Evaluation

3.12.15.1 Specific Weight Measurement

The specific gravity of acid is measured using a device at atmospheric pressure and a temperature of 25 °C. The specific gravity is used to calculate the percentage purity of the acid.

3.12.15.2 Measure the Amount of Iron Ions

Pour 10 ml of the acid solution into a beaker and adjust the pH to 4 using a pure strong base free of metal ions. Increase the volume of the solution to 100 ml with double-distilled water. Using a spectrophotometer, measure both the total iron content and the iron(II) ions. Finally, multiply the values obtained by the dilution factor (10) and report the results (note that the maximum permissible iron ion concentration is 8 ppm).

3.12.15.3 Percentage Purity Measurement

Take a defined volume of acid with a pipette and titrate a standard base in the presence of phenolphthalein reagent.

Note: Using the volume, specific gravity, and molecular weight of the hydrochloric acid, the volume, and the normality of reuse, calculate and report the percentage purity of the acid. The acceptable percentage by weight (concentration) for hydrochloric acid is 28–32%.

3.12.16 Acetic Acid Evaluation

3.12.16.1 Specific Gravity Measurement

The specific gravity of acid is measured using a device at atmospheric pressure and a temperature of 25 °C.

3.12.16.2 Measure the Amount of Iron Ions

Pour 10 ml of the acid solution into a beaker and adjust the pH to 4 using a pure strong base free of metal ions. Increase the volume of the solution to 100 ml with double-distilled water. Using a spectrophotometer, measure both the total iron content and the iron(II) ions. Finally, multiply the values obtained by the dilution factor (10) and report the results (note that the maximum permissible iron ion concentration is 8 ppm).

3.12.16.3 Percentage Purity Measurement

Take a defined volume of acid with a pipette and titrate a standard base in the presence of phenolphthalein reagent.

Note: Calculate and report the percentage purity of the acid using the specific gravity volume, the molecular weight of the acetic acid, the volume, and the recycling normality. The acceptable percentage by weight (concentration) for acetic acid is 15%.

3.12.17 Evaluation of Ammonium Bifluoride (ABF)

A specific quantity of ammonium bifluoride sample is weighed into a density measuring vessel (pycnometer) and then filled with an organic phase of a specific density (toluene). By measuring the weight of the organic phase added, its volume is calculated and by subtracting it from the volume of the pycnometer container, the volume of the ammonium bifluoride sample is obtained. Finally, determine the density of the sample by dividing the weight of the sample by the calculated volume.

3.12.17.1 Percentage Purity Measurement

Place a predetermined weight of ammonium bifluoride sample in a container and titrate with a standard base in the presence of methyl red reagent.

Note: Calculate and report the percentage purity of the ammonium bifluoride sample from the weight of the sample taken and the volume and normality of the reagent used for the titration.

3.12.17.2 Metal Ion Measurement

Dissolve a given weight of ammonium bifluoride in 100 ml of double-distilled water. Measure the concentrations of calcium, magnesium and iron ions using a spectrophotometer. Express the results as percentages by weight.

Note: It is recommended to use a plastic container for all stages of the evaluation of ammonium bifluoride.

3.12.17.3 Melting Point Measurement

Fill the capillary tube of the melting point measuring device with ammonium bifluoride particles. Place the mentioned tube in the measuring center and turn on the device. Record the melting temperature of the sample to ensure the result and repeat the test 2 or 3 times to increase the accuracy.

3.12.18 Benzoic Acid Evaluation

3.12.18.1 Specific Gravity Measurement

Weigh a portion of the benzoic acid sample into a density measuring vessel (pycnometer) of predetermined weight and volume. Fill the pycnometer with distilled water. Measure the weight of the added water and determine its density. From these data, calculate the volume of the water. Subtract this volume from the total volume of the pycnometer to obtain the volume of the benzoic acid sample. Finally, determine the density of the benzoic acid.

3.12.18.2 Percentage Purity Measurement

Dissolve a predetermined weight of benzoic acid sample in ethyl alcohol and titrate with standard potassium hydroxide in the presence of phenolphthalein reagent.

Note: Calculate and report the percentage purity of benzoic acid from the volume and normality of the potassium hydroxide consumed and the weight of the benzoic acid sample.

3.12.18.3 Melting Point Measurement

Fill the capillary tube of the melting point meter with benzoic acid particles. Place the tube in the measuring centre and activate the instrument. Record the melting temperature of the sample. To validate the results and improve accuracy, consider performing the test 2 or 3 times.

3.12.18.4 Particle Size Determination

To determine the particle size of the benzoic acid sample, pass 100 g of the sample through sieves of 75 μm (μm), 150 μm, 425 μm, 700 μm, 1 mm (mm), and 4 mm. Finally, report the percentage of particles passing through each sieve.

3.12.18.5 Water Solubility Measurement

Weigh a defined quantity of the sample into a 250 ml beaker and add 100 ml of distilled water. Place the beaker and its contents on a magnetic stirrer set at 400 rpm. After 1 h, filter the solution in the beaker and calculate the percentage solubility of the benzoic acid sample in water.

3.12.19 Asphaltene Solvent Evaluation

3.12.19.1 Static Test

First, set the temperature of the water bath to 70 °C. Next, obtain a 425 μm sieve and weigh 5–10 g of asphaltene deposits on the same sieve. Place the sieve with the sediment in a suitable container with 100 ml of the chosen solvent. Ensure that the sediment is completely immersed in the solvent and that a magnetic stirring rod can be inserted. Once the water bath temperature has been confirmed, place the assembly (solvent container, magnetic stirrer, sieve, and sediment) in the hot water bath. Set the stirrer at 400 rpm. After one hour, remove the assembly from the water bath and allow it to air dry at room temperature. Then weigh the sample and determine the percentage efficiency of the selected solvent based on the initial weight. Repeat the test under the same conditions after 3 h.

Note: The minimum acceptable efficiency for the asphaltene solvent over a 3-h dynamic period is 97%.

3.13 Coreflood Tests

3.13.1 Introduction

A coreflood device is used to replicate the conditions of a reservoir and well environment in a laboratory environment. Coreflood is a valuable tool for simulating processes such as acidizing and investigating formation damage. Throughout the testing process, various parameters such as initial and reverse permeability, as well as the penetration rate of the acidizing fluid into the core sample, are carefully examined. The practice of coreflood of cores is a distinctive approach aimed at investigating the effects of matrix acidizing on different rock types. The primary objective of acid coreflood is to explore different conditions and analyse their effects on different rock formations after acidizing. Hydrochloric acid (HCl) is usually the acid of choice for treating carbonate formations, as it reacts readily with carbonate minerals. When dealing with carbonate formations, it is essential to consider both the damage incurred and the properties of the rock when designing the treatment. In carbonate rock, stimulation of the effective matrix results in the formation of permeable channels known as "wormholes". These wormholes are a direct result of the acid-mineral reaction. The structure of these wormholes is significantly influenced by the injection rate and the fluid properties of the rock. At lower injection rates, wormholes can lead to complete dissolution from the injection surface. Conversely, at higher injection rates, more uniform dissolution and branching wormholes may occur. At moderate flow rates, a dominant wormhole structure typically develops.

Coreflood is a laboratory technique used to replicate the conditions of a reservoir and well environment. The equipment used assesses the potential for formation damage. Figs. 3.71, 3.72, 3.73, 3.74, and 3.75 illustrate the FDS350 coreflood tool and its components.

Fig. 3.71 Air compressor device

Fig. 3.72 Flooding device FDS350

Fig. 3.73 Device manual valves

Figure 3.76 shows a schematic of the connection of the various components, which is provided for a better understanding of the device and how to work with it.

3.14 Coreflood with Acid

3.14.1 Matrix Acidizing Test Details

Coreflood with acid is a unique method used to study the effects of matrix acidizing on different rock types. The primary objective of this procedure is to investigate different conditions and their influence on different rock formations after acidizing. Hydrochloric acid (HCl) is commonly used to acidizing carbonate formations due to its ready reactivity with carbonate minerals. When dealing with carbonate formations,

Fig. 3.74 Nitrogen gas capsule

Fig. 3.75 Valves for confining pressure

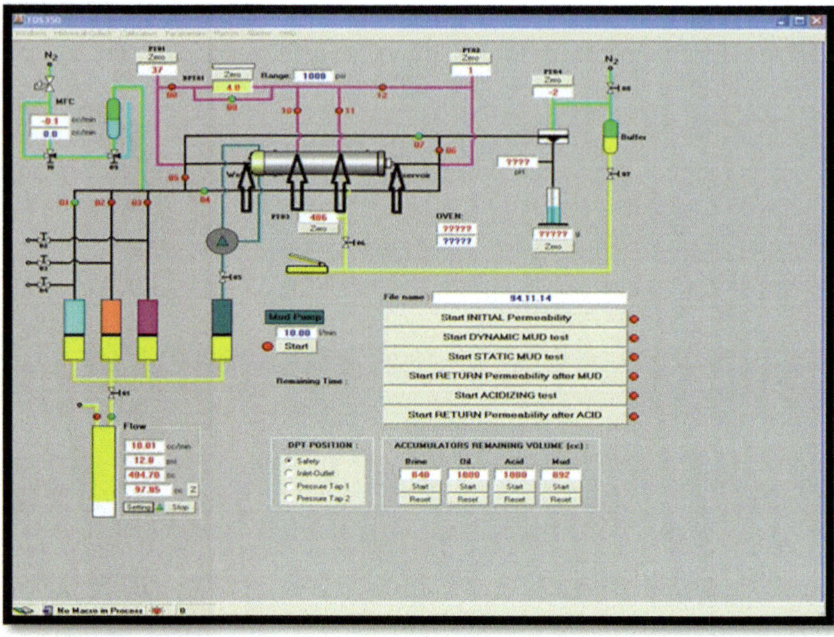

Fig. 3.76 Device map and manual and pneumatic valves

it is essential to consider both the damage incurred and the properties of the rock when designing the treatment.

In carbonate rock, stimulation of the active matrix results in the formation of permeable channels known as "wormholes". These wormholes are the direct result of the acid-mineral reaction. The structure of these wormholes is strongly influenced by the injection rate and the fluid properties of the rock. At lower injection rates, the wormhole structure can result in complete dissolution from the inlet surface. Conversely, at higher injection rates, a more uniform dissolution process can occur, leading to the formation of branched wormholes. At intermediate flow rates, a dominant wormhole structure typically develops.

3.14.2 Matrix Acidizing Method

The general method of matrix acidizing is as follows:

1. Cutting and preparation of core samples.
2. Saturate core samples with formation water.
3. Preparation of acid mixture and additives
4. Fill pumps with hydraulic oil and accumulators with acid and water.
5. Setting the input and output levels and placing the core in the holder.
6. Set the desired back pressure from the nitrogen source.

7. Pump the fluids from the inside of the core sample in the correct sequence. If temperature control is required (to simulate the reservoir environment), the oven is activated and acid pumping begins.
8. Switch on the pump, close the inlet valve, allow the return pressure to reach zero, and apply the required limiting pressure.
9. Dismantle the equipment and clean all parts that have been in contact with the fluid.

3.14.3 Acid Preparation

Different concentrations and strengths of acid must be prepared for use in different types of matrix acidizing experiments. When acidizing sandstone core samples, a mixture of HF and HCl or weaker acids is used. The concentration of these acids may vary depending on the desired effect on mineral dissolution or the degree of damage removal. For carbonate formations, which form the core of the reservoir, hydrochloric acid (HCl) is the preferred choice for stimulation. The specific concentration required can vary depending on the nature of the rock. Acid mixtures used in core flooding experiments include 15% HCl, 15% HCl + 10% HAc, 15% HCl + MeOH, and 7.5% HCl + MeOH. These acids are combined with corrosion inhibitors, corrosion inhibitor intensifiers, iron controllers, and surface tension reducers in percentages of 1.5, 1.5, 1, and 2 respectively.

3.14.4 Determining the Optimal Flow Rate

Acids naturally tend to flow into larger pores due to the lower flow resistance in such pores. However, it's important to note that larger pores are not always connected to other larger pores within the rock structure. Many researchers have conducted studies to investigate the factors that influence optimal conditions through core flooding experiments. These factors include reservoir temperature, hydrochloric acid concentration, and the type of carbonate rock involved. Experimental results indicate that increasing the temperature can increase both the optimal acid flux and the corresponding optimal pore volume. On the other hand, a higher acid concentration results in a higher acid flux but a lower optimal pore volume.

In order to determine the optimum flow rate, all influencing conditions must be considered, including core permeability, porosity, time, acid viscosity, and pH.

3.14.5 Coreflood Test Steps

The first step in the test is to inject formation water to measure the permeability of the core. To ensure that the carbon dioxide produced dissolves in the fluid, a back pressure regulator is used to maintain pressure at the outlet of the core. A core sample with specific porosity and permeability characteristics is used for the test. The acid is then injected into the core at an optimum flow rate. As the wormholes begin to form, the pressure drop across the core gradually decreases. When the pressure drop reaches almost zero, it indicates that the wormholes have left the core.

3.15 Questions

Question 3.1 Consider an acidizing reaction between acetic acid (CH_3COOH) and dolomite ($CaMg(CO_3)_2$) represented by the following chemical equation:

$$4CH_3COOH + MgCa(CO_3)_2 \leftrightarrows Mg(CH_3COO_2)_2 + Ca(CH_3COO_2)_2 + 2H_2O + 2CO_2$$

a. Use Eq. 3.5 to calculate the mass dissolution power (β) for this reaction. Given that $\vartheta_{mineral} = 1$, $\vartheta_{acid} = 4$, $MW_{mineral} = 184g/mol$, $MW_{acid} = 60g/mol$, and the mass of acid in the solution is 80 g.
b. Determine the mass of calcite that can be dissolved if the total mass of the solution is 200 g.
c. Using the calculated mass dissolution power from part A, find the volumetric dissolution power (X) using Eq. 3.6. Assume the density of the acid solution ($\rho_{acidsolution}$) is 1.2 g/mL and the density of calcite ($\rho_{mineral}$) is 2.85 g/cm^3.
d. The rate of acid formation (R_A) for the reaction is given by Eq. 3.7. For the reaction $E_f = 0.002\frac{molA}{m^2-s-(\frac{molA}{m^3})^{\alpha}}$, $C_A = 0.02$ mol/m^3, and $\alpha = 0.8$. Calculate R_A using these parameters.
e. Determine the equilibrium constant (K_{eq}) for the given reaction based on the concentrations of the reactions and products. Given that $[CH_3COOH] = 0.03$ M, $[MgCa(CO_3)_2] = 0.02M$, $Mg(CH_3COO_2)_2 = Ca(CH_3COO_2)_2 = 0.01M$, and $H_2O = CO_2 = 0.02$ M.

Question 3.2 Examine the following scenario and answer the questions.

Imagine you have a solution with a pH of 5.5. Discuss the key aspects of this solution, considering its acidity or basicity. Additionally, explain how the temperature might influence the pH measurement. If you were to add a small amount of a strong base to this solution, how would the pH and pOH change? Finally, relate the principles of buffer solutions to this scenario and discuss how the addition of a weak acid might impact the solution's pH.

Question 3.3

A. The cross-section of a rock sample for an acid-rock test is photographed, revealing a diameter of 7 cm and a thickness of 3 cm. determine the area of the reaction surface.

B. The acid-dolomite chemical reaction is represented by:

$$2HCl + CaCO_3 \rightarrow CaCl_2 + CO_2 + H_2O$$

Using the given equations, calculate the initial concentration of the acid solution with the following parameters:

$$\rho_0 = 1.5\,\frac{g}{cm^3},\quad C_{NaOH} = \frac{0.1\ mol}{L},\quad V_{NaOH-1} = 50\ mL,\quad m_{HCl-1} = 30\ g$$

C. Using the given equations, determine the concentration of acid after the reaction with the following parameters:

$$\rho_1 = 1.2\,\frac{g}{cm^3},\quad C_{NaOH} = \frac{0.1\ mol}{L},\quad V_{NaOH-2} = 25\ mL,\quad m_{HCl-2} = 15\ g$$

D. Calculate the dynamic acid-rock reaction rate with the following parameters:

$$t_1 = 400\ s,\ t_2 = 200\ s,\ V = 200\ mL$$

References

Ali, S. A., Kalfayan, L., & Montgomery, C. (2016). *Acid stimulation*. Monograph Series, Society of Petroleum Engineers.

Alkattan, M., Oelkers, E. H., Dandurand, J. L., & Schott, J. (1998). An experimental study of calcite and limestone dissolution rates as a function of pH from −1 to 3 and temperature from 25 to 80 C. *Chemical Geology, 151*(1-4), 199–214.

Ansari, K. R., et al. (2020a). *Corrosion inhibitors for acidizing process in oil and gas sectors* (pp. 151–176).

Ansari, K. R., Chauhan, D. S., Singh, A., Saji, V. S., & Quraishi, M. A. (2020b). *Corrosion inhibitors for acidizing process in oil and gas sectors* (pp. 151–176). Corrosion Inhibitors in the Oil and Gas Industry.

Assem, A.I. (2013). *Formation damage due to iron precipitation during matrix acidizing treatments of carbonate reservoirs and ways to minimize it using chelating agents.*

Assem, A. I., Nasr-El-Din, H. A., & De Wolf, C. A. (2013). Formation damage due to iron precipitation in carbonate rocks. *Paper presented at the SPE European formation damage conference and exhibition.*

Berry, J. D., Neeson, M. J., Dagastine, R. R., Chan, D. Y., & Tabor, R. F. (2015). Measurement of surface and interfacial tension using pendant drop tensiometry. *Journal of Colloid and Interface Science, 454*, 226–237.

Chacon, O. G., & Pournik, M. (2022). Matrix acidizing in carbonate formations. *Processes, 10*(1), 174.

Da Motta, E., Miranda, C., Anjos, S., Ribeiro, J., & Chaves, E. (1996). Acidizing wells with acetic/HF acid mixtures to minimize cement dissolution. *SPE international conference and exhibition on formation damage control.*

Da Motta, E., Quiroga, M., Aragão, A., & Pereira, A. (1998). Acidizing gas wells in the Merluza field using an acetic/formic acid mixture and foam pigs. *SPE international conference and exhibition on formation damage control.*

De Caprariis, B., Stoller, M., Chianesea, A., & Verdone, N. (2015). CFD model of a spinning disk reactor for nanoparticle production. *Chemical Engineering, 43*, 757–762.

Economides, M. J., & Nolte, K. G. (1989). *Reservoir stimulation* (Vol. 2). Prentice Hall Englewood Cliffs, NJ.

Fan, Y., Peng, H., Chen, G., Peng, J., Han, H., Qin, Y., Wang, L., & Liu, D. (2023). Experimental study of the influences of different factors on the acid-rock reaction rate of carbonate rocks. *Journal of Energy Storage, 63*, 107064.

Garrouch, A. A., & Jennings, A. R. (2017). A contemporary approach to carbonate matrix acidizing. *Journal of Petroleum Science and Engineering, 158*, 129–143.

Gdanski, R. (1999). A fundamentally new model of acid wormholing in carbonates. In *SPE European Formation Damage Conference and Exhibition* (pp. SPE-54719). SPE.

Hall, B., & Dill, W. (1988). Iron control additives for limestone and sandstone acidizing of sweet and sour wells. *SPE international conference and exhibition on formation damage control.*

Harris, F. (1961). Applications of acetic acid to well completion, stimulation and reconditioning. *Journal of Petroleum Technology, 13*(07), 637–639.

Harris, O., Hendrickson, A., & Coulter, A. (1966). High-concentration hydrochloric acid aids stimulation results in carbonate formations. *Journal of Petroleum Technology, 18*(10), 1291–1296.

Herman, J. S., & White, W. B. (1985). Dissolution kinetics of dolomite: Effects of lithology and fluid flow velocity. *Geochimica et Cosmochimica Acta, 49*(10), 2017-2026.

Huang, T., Ostensen, L., & Hill, A. (2000). Carbonate matrix acidizing with acetic acid. *SPE international conference and exhibition on formation damage control.*

Kalatehno, J. M., Khamehchi, E., Abbasi, A., & Khaleghi, M. R. (2023). A novel approach to determining appropriate additive concentrations for stimulation of gas carbonate reservoirs. *Results in Engineering, 12*, 101440.

Khalil, R., Emadi, H., & Altawati, F. (2021). Investigating the effect of matrix acidizing injection pressure on carbonate-rich Marcellus shale core samples: An experimental study. *Journal of Petroleum Exploration and Production, 11*, 725–734.

King, G., & Lee, R. (1988). Adsorption and chlorination of mutual solvents used in acidizing. *SPE Production Engineering, 3*(02), 205–209.

Lund, K., Fogler, H. S., McCune, C. C., & Ault, J. W. (1975). Acidization—Ii. The dissolution of calcite in hydrochloric acid. *Chemical Engineering Science, 30*(8), 825-835

Mirvakili, A., Rahimpour, M., & Jahanmiri, A. (2012). Experimental study of iron-control agents selection for high temperature sour gas and oil wells acidizing process. *Chemical Engineering Research and Design, 90*(11), 1823–1833.

Qiu, X., & Chang, F. (2011). Raman spectroscopy yields new discovery of reaction of acetic acid with calcite. *SPE Kingdom of Saudi Arabia annual technical symposium and exhibition.*

Rashed, M. K., Abdulbari, H. A., Salled, M., & Ismail, M. H. (2016). Rotating disc apparatus: Types, developments and future applications. *Modern Applied Science, 10*(8), 198–229.

Ratnakar, R., Kalia, N., & Balakotaiah, V. (2012). Carbonate matrix acidizing with gelled acids: An experiment-based modeling study. *SPE international production and operations conference and exhibition.*

Smith, C., Crowe, C., & Nolan, T. (1969). Secondary deposition of iron compounds following acidizing treatments. *Journal of Petroleum Technology, 21*(09), 1121–1129.

Smith, C. F., & Hendrickson, A. R. (1965). Hydrofluoric acid stimulation of sandstone reservoirs. *Journal of Petroleum Technology, 17*(02), 215–222.

Taylor, K., & Nasr-El-Din, H. (2009). Measurement of acid reaction rates with the rotating disk apparatus. *Journal of Canadian Petroleum Technology, 48*(06), 66–70.

Taylor, K. C., Nasr-El-Din, H. A., & Mehta, S. (2006). Anomalous acid reaction rates in carbonate reservoir rocks. *SPE Journal, 11*(04), 488–496.

Toseef, A., Beaman, D. J., & Birou, P. (2010). Viscoelastic surfactant diversion: An effective way to acidize low-temperature wells. *Abu Dhabi international petroleum exhibition and conference.*

Witkowski, A., Rusin, A., Majkut, M., & Stolecka, K. (2014). The analysis of pipeline transportation process for CO_2 captured from reference coal-fired 900 MW power plant to sequestration region. *Chemical and Process Engineering, 35*(4), 497–514.

Chapter 4
Analyzing and Contrasting Laboratory Results with Commercial Software

4.1 Introduction

In this chapter, we have simulated the acidizing operation of well No. 15 using commercial software. To begin, we provide the necessary information about the reservoir and the well, which the software requires. Following that, we introduce the software itself along with its features. Subsequently, we compare various acidizing programs and present the most optimal mode from both an economic and technical perspective. Finally, we present the results and software-generated graphs for the chosen optimal mode.

This software proves to be well-suited for executing acid processes. Acid treatment stands as a crucial method to reconnect the formation and the well column post-drilling. The software is capable of performing diverse calculations and simulating acid operations through the incorporation of characteristics such as well completion type, well specifications, wall and core tubes, reservoir fluid type, reservoir data, and its corresponding layers, as well as reservoir rock lithology and history. Every chemical compound employed in acid operations is encompassed within this program, enabling experts to precisely simulate their desired compounds. Through its understanding of acid corrosiveness and its impact on formation characteristics, the program can predict and compute the formation's vulnerability degree, along with potential risks associated with the process. Notably, the software exhibits exceptional accuracy when simulating carbonate reservoirs, being meticulously designed with a focus on acidizing operations, which ensures the integration of features that streamline the simulation process. Engineers specializing in oil and gas can effectively conduct acidizing procedures within oil and gas reservoirs through the use of this software. The utilization of this program facilitates secure testing within a virtual environment, thus minimizing costs and potential risks, before the execution of physical acidizing processes within operational contexts, ultimately leading to more informed conclusions.

© The Author(s), under exclusive license to Springer Nature Switzerland AG 2024

E. Khamehchi et al., *Applied Matrix Acidizing of Carbonate Reservoir*,
Petroleum Engineering, https://doi.org/10.1007/978-3-031-58281-3_4

The obstacles impeding the realization of the software's simulation results in acidification operations encompass:

1. Inadequate adherence to the predefined conditions within the software for performing the acid operation.
2. Discrepancies between the acid utilized in the operation and the acid specified within the software.
3. The presence of minerals existing in the reservoir rock that are not accounted for in the software. These minerals may exist in practice, and their omission leads to the formation of wormholes and channels, diverting the acid and hindering its penetration to the intended depth.

4.2 Required Information

Vital information required includes reservoir pressure and well depth, two of the most crucial parameters. Additional details pertaining to this well, encompassing completion methodology, wall tube, and cerebral tube diameters, are provided in the list below. Moreover, the completion design for this well takes the form of an open hole.

- Tubing and casing specifications comprise the following:
  - Burst pressure(psi),
  - Collapse pressure(psi),
  - Pipe grade (D),
  - Weight (pounds per foot),
  - Inner diameter (inches),
  - Outer diameter (inches),
  - Depth (meters).

Data concerning the reservoir fluid is outlined in Fig. 4.1. Additionally, requisite information such as oil, gas, and produced water percentages, reservoir volume, and reservoir radius, among other details, are mandated inputs for the software.

- Reservoir Information:
  - Surface temperature (Fahrenheit),
  - Well bottom temperature (Fahrenheit),
  - Radius of the well (ft),
  - Radius of the reservoir (meters),
  - Reservoir area (Km^2),
  - Reservoir pressure (psi).
- Saturation Percentage:
  - Water,
  - Gas,

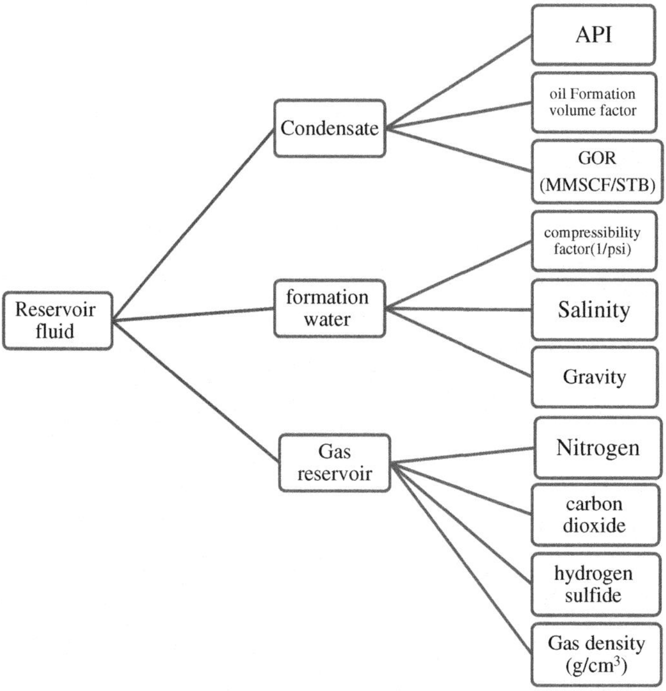

Fig. 4.1 Necessary reservoir fluid information for the software

– Oil.

Depth information for the target layer, intended for acidizing, is necessary. Additionally, details about this layer, including permeability, porosity, flow capacity, pressure gradient, and layer pressure, can be found in Fig. 4.2. If the well is open, there is no need to provide information about perforation.

The lithology associated with each layer plays a pivotal role in acidizing operations, significantly influencing acid selection and the sediments produced post-acid application. Essential data includes mineral percentages like calcite, quartz, dolomite, and smectite/elite. By inputting well completion details, drilling specifics, reservoir particulars, formation damage insights, layer attributes, and acid treatment plans, the software models the well's post-acid treatment condition, providing estimations of formation damage incurred by the operation.

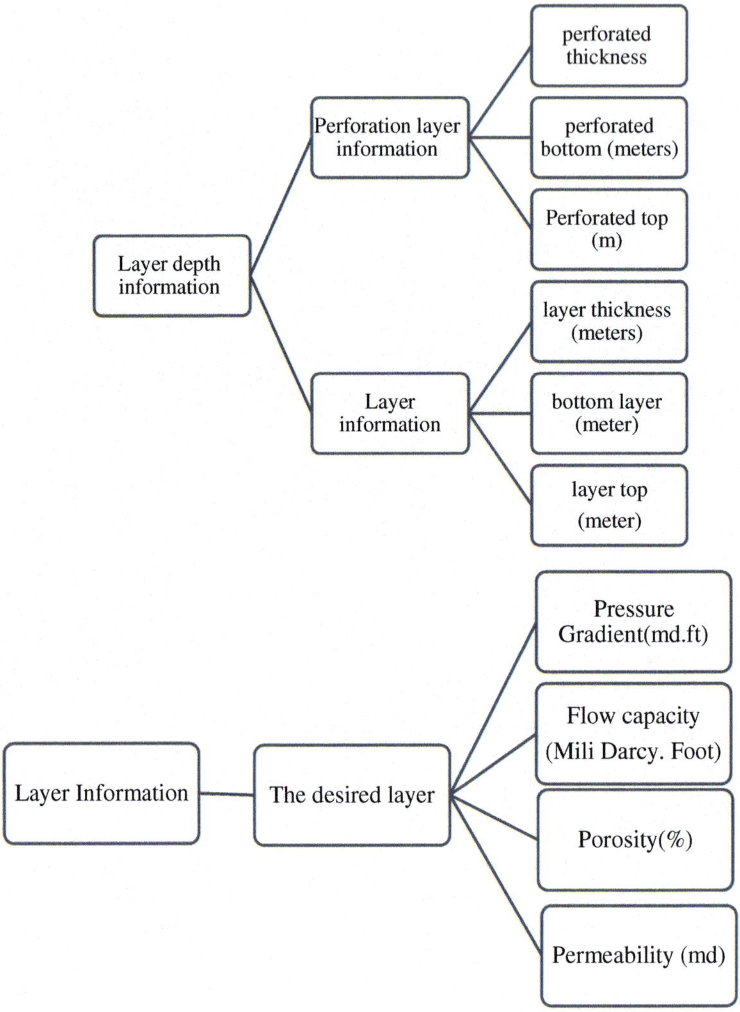

Fig. 4.2 Depth and flow information for the specified layer within the software

4.3 Test Design Method in Software

To achieve specific criteria within an experiment, it becomes essential to define all parameters influencing a process through careful experimental design. Thus, this section employs factorial design to simulate the acidizing operation. The objective behind simulating the operation for the intended well is to determine the optimal acidizing program. Following the execution of various models, the software identifies the most advantageous matrix acidizing program, considering economic and efficiency factors, and designates it as the optimal acid. Table 4.1 outlines distinct

Table 4.1 Designing simulation experiments for acidizing operations

	Acid type	Acid volume	Acid flow
Levels	c	b	a
Total	c × b × a		

factors and levels for conducting factorial simulation. Based on this table, diverse acid types are employed across varying volumes and flow rates. Notably, the chosen acids align with commonly used ones in oil and gas well acidizing. Additionally, the software recommends an optimal flow rate and volume, which are advised to be set three values lower for an even more optimized selection concerning these two parameters.

4.3.1 *Rationale Behind the Selection of Experiment Design Levels*

Hydrochloric acid stands as the prevailing choice for carbonate reservoir acidizing due to its remarkable solvent capabilities. However, its utility is curbed by two limitations: its elevated reaction rate and equipment corrosion. To regulate the reaction rate, a blend of hydrochloric acid and organic acids is employed. This amalgamation optimizes wormhole penetration depth. Laboratory tests have revealed that the optimal concentration for hydrochloric acid is 15%. Additionally, recognizing that a weaker acid achieves greater penetration, laboratory tests also assessed hydrochloric acid at a concentration of 7.5%.

As per the study by Jafar Pour et al., acidizing carbonate reservoirs necessitates the use of a blend comprising 8–13% acetic acid and 5–9% formic acid alongside hydrochloric acid. Following tests on nine different combinations, they identified the combination of 15% hydrochloric acid and 13% acetic acid as the most effective. Omer Shafiq et al.'s experiments demonstrated that a blend of 7.5% hydrochloric acid and 2.5% acetic acid achieves superior penetration compared to hydrochloric acid alone, leading to increased permeability. In 2008, Chang and colleagues investigated combinations of 15% hydrochloric acid with 10% acetic acid and 15% hydrochloric acid with 9% formic acid. Their research indicated that these compounds not only diminish corrosion but also lead to reduced acid loss. In light of reservoir-specific attributes and the aim for substantial penetration, exploration also encompassed the combination of 7.5% hydrochloric acid with 13% acetic acid, along with 7.5% hydrochloric acid with 9% formic acid.

Regarding injection rates, the software proposes a maximum rate, while other rates stand at 75, 50, and 25% of the maximum. Fourteen illustrative software-generated outcomes are presented in Table 4.2.

Table 4.2 Outcomes derived from simulated acidizing operation performances

Number	Acid type	Permeability (md)	Injection flow (bbl/min)	The volume of injected acid (bbl)	Skin factor	Injection duration (hours)	Cost (Doller)	ΔQ (million standard cubic feet per day)
1	HCl 7/5%	1	3.1	500	−1.41	13.9	866.67	35.151
2	HCl 7/5%	1	3.1	375	−1.41	13.2	650	35.151
3	HCl 15%	1	2.32	375	−1.13	13.9	1300	22.595
4	HCl 15%	1	2.32	250	−1.13	13	866.67	22.595
5	HCl 7/5% + Acetic2/5%	1	2.32	500	−1.31	14.8	104.67	29.875
6	HCl 7/5% + Acetic2/5%	1	2.32	375	−1.31	13.9	785	29.875
7	HCl 7/5% + Acetic 13%	1	3.1	125	−1.33	11.9	724.17	30.949
8	HCl 7/5% + Acetic 13%	1	2.32	500	−1.26	14.8	289.67	27.711
9	HCl 15% + Acetic 10%	1	3.1	375	−1.08	13.2	2230	20.705
10	HCl 15% + Acetic 10%	1	3.1	250	−1.08	12.5	148.67	20.705
11	HCl 15% + Acetic 13%	1	3.1	500	−1.05	13.9	350.33	19.701
12	HCl 15% + Acetic 13%	1	3.1	375	−1.05	13.2	2627.5	19.701

(continued)

Table 4.2 (continued)

Number	Acid type	Permeability (md)	Injection flow (bbl/min)	The volume of injected acid (bbl)	Skin factor	Injection duration (hours)	Cost (Doller)	ΔQ (million standard cubic feet per day)
13	HCl 15% + Formic 9%	1	1.55	250	−0.55	13.9	399.17	7.653
14	HCl 15% + Formic 9%	1	1.55	125	−0.39	12.5	199.08	5.246

4.4 Selecting the Rate and Volume of Acid Injection

Initially, the rate and volume derived from inputting the necessary well and reservoir information into the software are treated as the maximum rate and volume, while other values are set at 75, 50, and 25% of the maximum value, respectively.

4.5 Discussion

4.5.1 Optimal Acid Selection

Considering the significant occurrence of skin factors less than negative ones, the primary criterion for optimal model selection prioritizes those with skin factors below negative ones. Consequently, all acidizing programs featuring a skin factor greater than a negative one are excluded. Subsequently, the evaluation turns to models that share the same skin factor post-acidizing. In this scenario, models with equivalent skin factors, but higher costs are eliminated. Further scrutiny involves models sharing the same cost but exhibiting different skin factors. Within this context, the acid program with the higher skin factor is removed.

Continuing this iterative process and examining simulation outcomes, diverse models are progressively eliminated based on criteria including skin factor, acid working duration, and the cost of acid consumption. Ultimately, the acid or acids fulfilling these conditions emerge as the final selections.

4.5.2 *Information and Outcomes from the Software*

4.5.2.1 Hole Length by Volume Diagram

Figures 4.3, 4.4, 4.5 and 4.6 illustrate the length of wormholes generated at each stage per volume for the following acid compositions: 7.5% HCl, 7.5% HCl + 2.5% HAc, 15% HCl, and 15% HCl + 13% HAc.

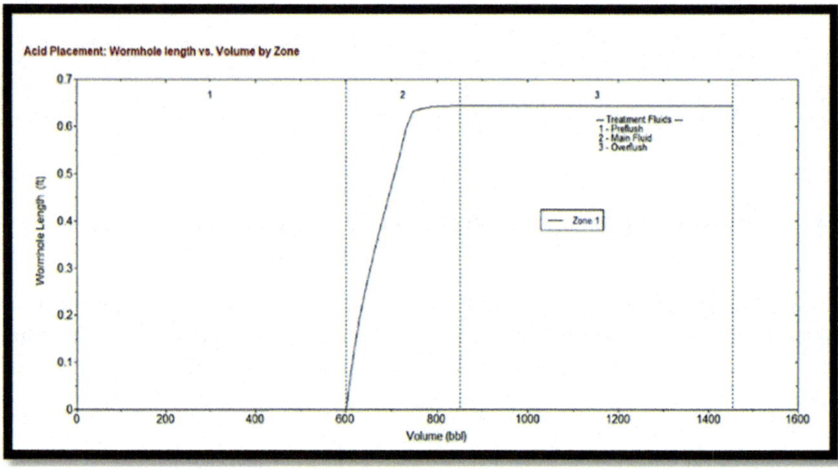

Fig. 4.3 Plot of pore length by volume for 7.5% HCl acid

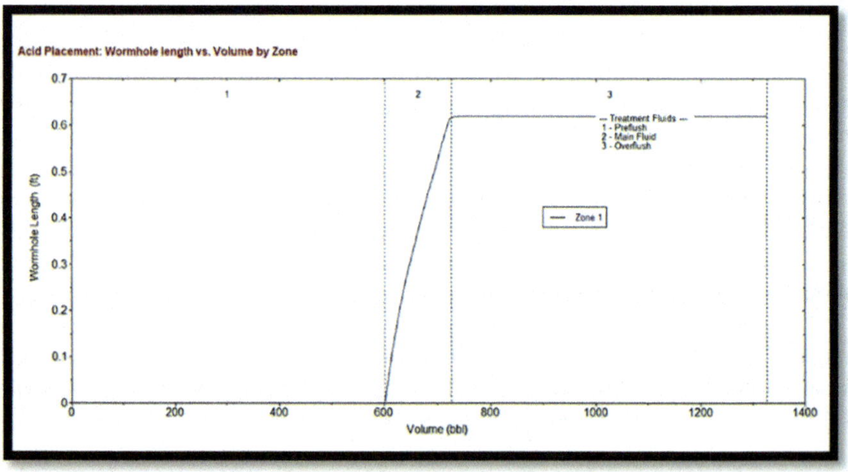

Fig. 4.4 Plot of pore length by volume for 7.5% HCl + 2.5% HAc acid

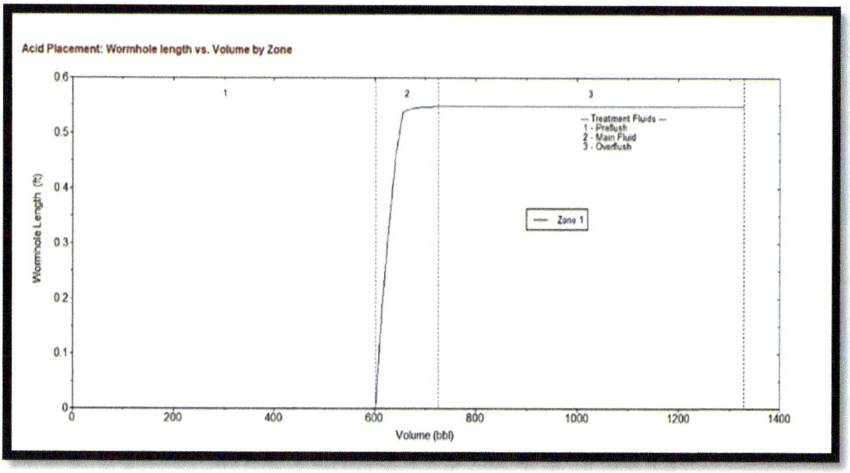

Fig. 4.5 Plot of pore length versus volume for 15% HCl acid

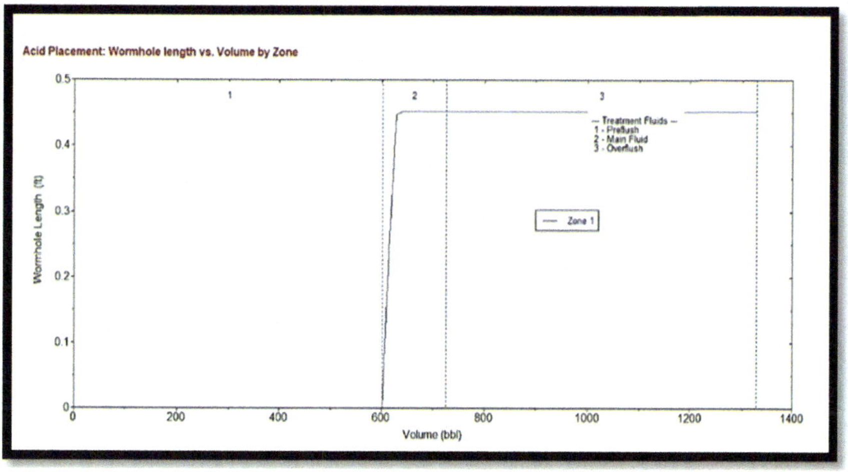

Fig. 4.6 Plot of pore length by volume for 15% HCl + 13% HAc acid

Based on Figs. 4.3, 4.4, 4.5 and 4.6, a preliminary flush fluid consisting of 6 barrels was introduced into the system. Following this, after 600 barrels of prewash fluid injection, the acid introduced began to react with the rock, carving a channel for itself, ultimately forming wormholes of up to 0.6 ft in length. Importantly, subsequent to the main acid injection, the post-acid flush fluid did not extend the cavity's length, culminating in a final cavity length of 0.6 ft.

Analyzing the simulation results and referencing Figs. 4.3, 4.4, 4.5 and 4.6, it becomes evident that 7.5% HCl demonstrates superior penetration due to its lower acid strength, leading to the creation of wormholes with more substantial depths.

Following this trend, 7.5% HCl + MeOH, 15% HCl, and 15% HCl + 13% HAc generated wormholes of varying lengths, in descending order. Dissolution tests revealed significantly lower dissolution for 7.5% HCl and 7.5% HCl + MeOH. However, simulation results underscore that weaker acids yield greater penetration depths.

4.5.2.2 The Diagram of Skin Factor by Volume

Figures 4.7, 4.8, 4.9 and 4.10 present the skin factor subsequent to each acidizing phase per acid consumption for the following acid compositions: 7.5% HCl, 7.5% HCl + 2.5% HAc, 15% HCl, and 15% HCl + 13% HAc. Furthermore, Table 3.5 provides the shell factor value obtained after acidizing through commercial software (Table 4.3).

In the acidizing process, the initial pre-flush fluid serves to create a favorable environment for rock reaction, displacing fluids that have dampened the reactivity of the rock. Following this, with the introduction of the primary fluid and subsequent after-flush fluid, the skin factor effect gradually diminishes until the injected acid is exhausted.

The pre-flash fluid, typically diesel, holds no reactivity with the rock and thus leaves the skin factor value unchanged. Subsequent acid injection leads to a decrease in the skin factor due to the rock-acid reaction and rock dissolution. The introduction of after-flash fluid primarily propels the acid further into the reservoir rock and has minimal impact on the skin factor.

Upon examining Figs. 4.7, 4.8, 4.9 and 4.10 and referring to Table 3.5, it's evident that the skin factor reduction is more prominent for 7.5% HCl and 7.5% HCl + 2.5% HAc compared to other acid compositions.

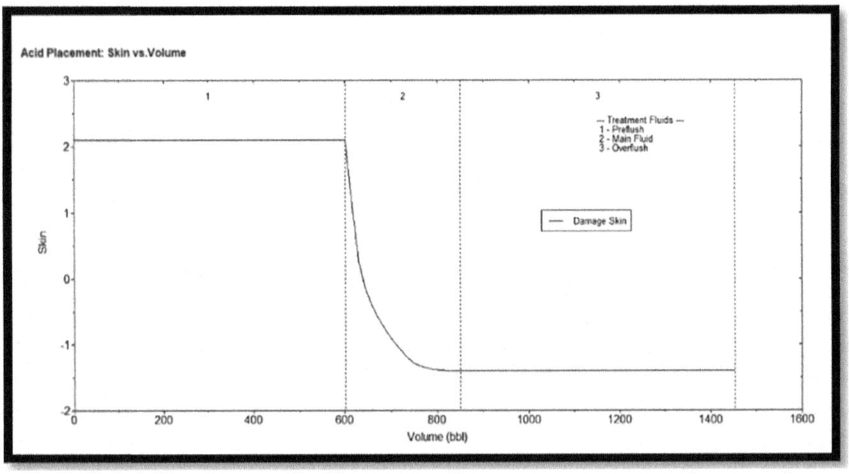

Fig. 4.7 Skin factor diagram by volume for 7.5% HCl acid

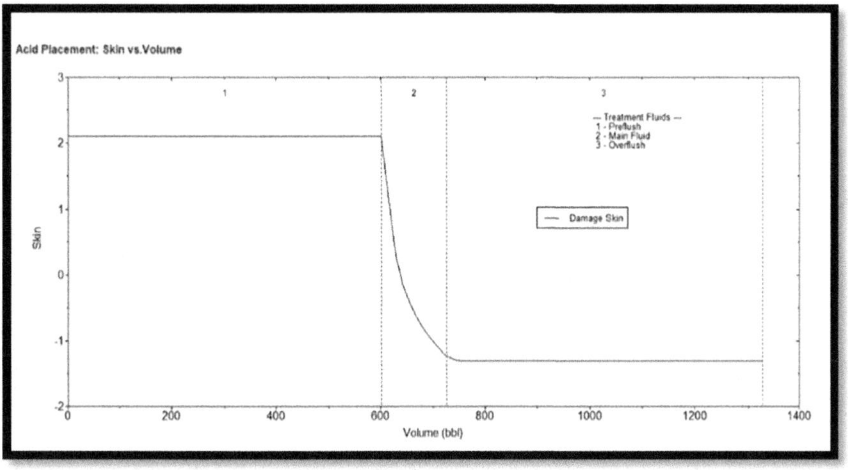

Fig. 4.8 Skin factor diagram by volume for HCl 7.5% + HAc 2.5

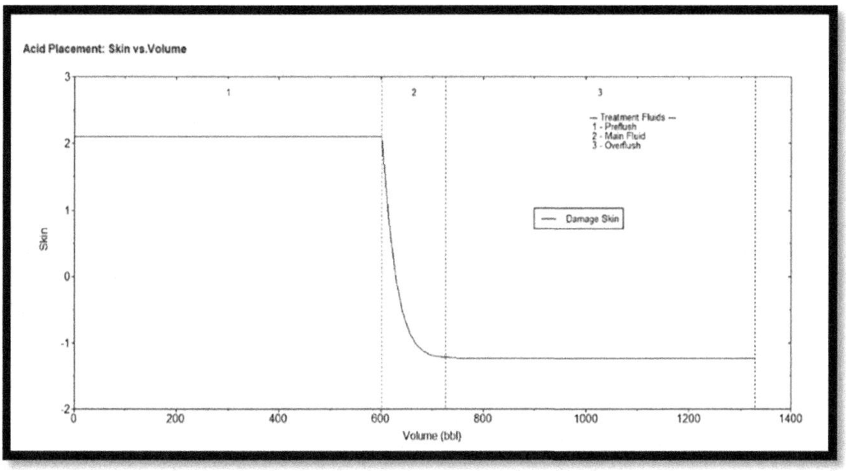

Fig. 4.9 Skin factor diagram by volume for HCl 15%

4.5.2.3 Chart of Acid Injection Rate Based on Acid Volume Consumed

Figures 4.11, 4.12, 4.13 and 4.14 illustrate the acid injection rate at each stage relative to the volume of injected acid for the acid compositions HCl 7.5%, HCl 7.5% + HAc 2.5%, HCl 15%, and HCl 15% + HAc 13%. Ultimately, the optimal flow rate for the desired acids is ascertained through the conducted investigations. The findings from these investigations are documented in Figs. 4.11, 4.12, 4.13 and 4.14.

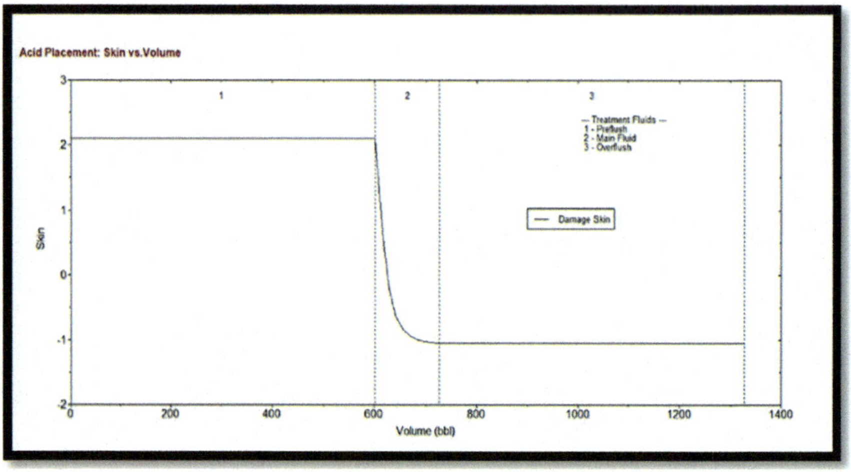

Fig. 4.10 Skin factor diagram by volume for HCl 15% + HAc 13%

Table 4.3 The outcomes of the skin factor values in the simulation

Acid	Skin factor
HCl 15% + HAc13%	−1.05
HCl 7/5%	−1.41
HCl 7.5% + HAc2.5%	−1.32
HCl 15%	−1.24

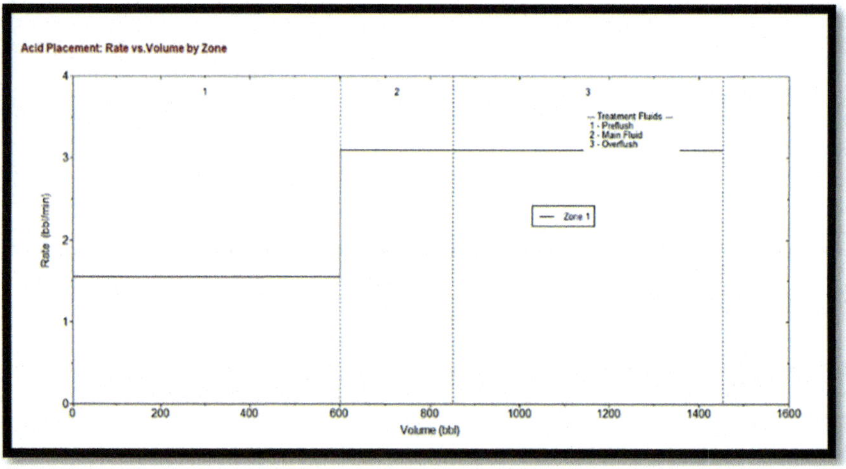

Fig. 4.11 Injection acid rate chart according to the volume of acid consumed for 7.5% HCl acid

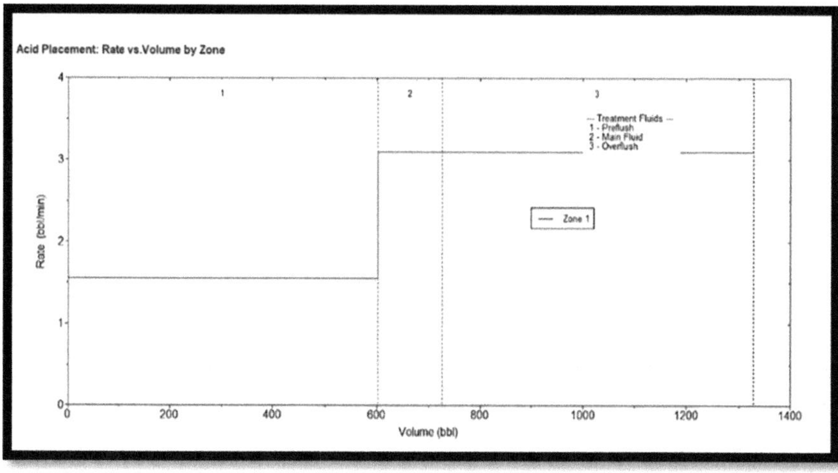

Fig. 4.12 Injection acid rate chart according to the volume of acid consumed for 7.5% HCl + 2.5% HAc acid

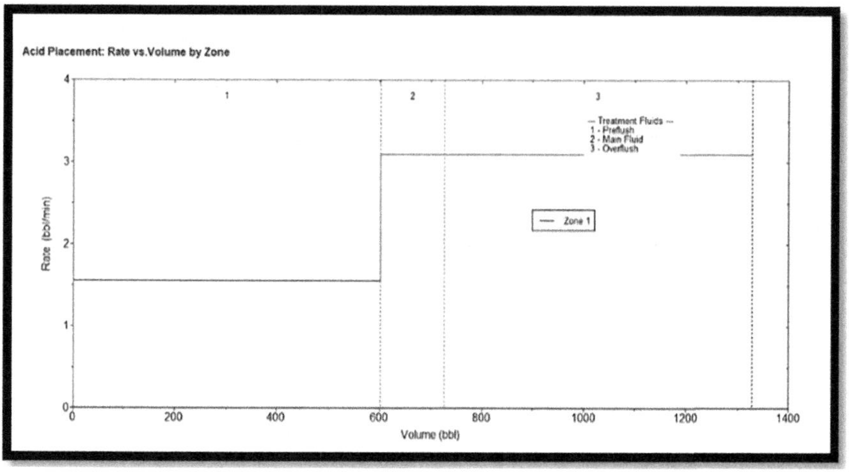

Fig. 4.13 Injection acid rate chart according to the volume of acid consumed for 15% HCl acid

4.5.2.4 Radius of Influence Diagram

Figures 4.15, 4.16, 4.17 and 4.18 depict the extent to which acidizing fluids penetrate the reservoir for the acids 7.5% HCl, 7.5% HCl + 2.5% HAc, 15% HCl, and 15% HCl + 13% HAc.

Figures 4.15, 4.16, 4.17 and 4.18 the volume of the pre-flash fluid and the primary acid is significantly less than that of the after-flash fluid. The pre-flash fluid serves the purpose of displacing the reservoir fluids to facilitate further acid-rock reactions.

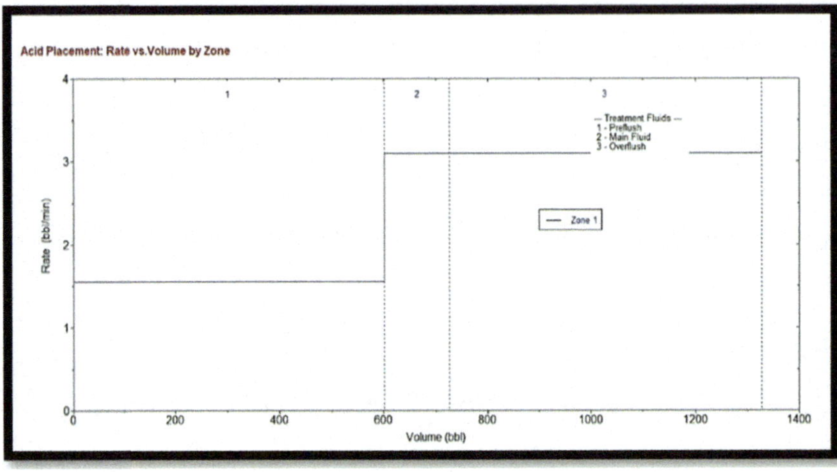

Fig. 4.14 Injection acid rate chart according to the volume of acid consumed for 15% HCl + 13% HAc acid

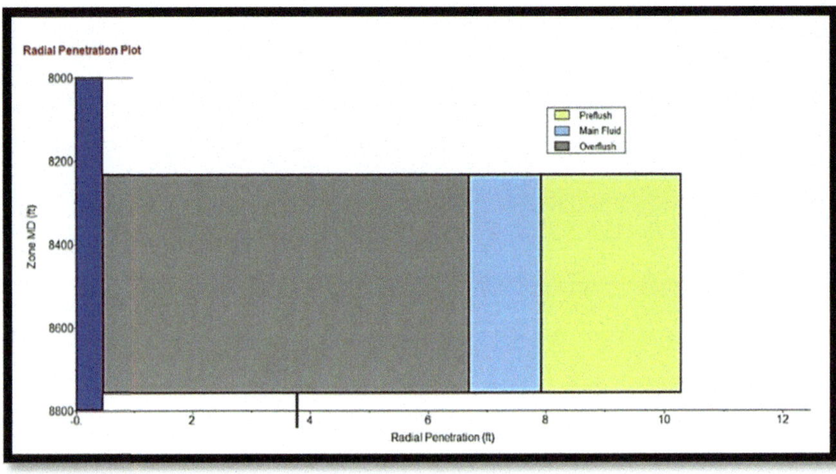

Fig. 4.15 Radius of penetration diagram for 7.5% HCl acid

Subsequently, the primary acid, following the pre-flush fluid, enhances permeability and mitigates existing damage. The after-flash fluid's role is to propel the initial acid forward and impact a broader area with greater effectiveness.

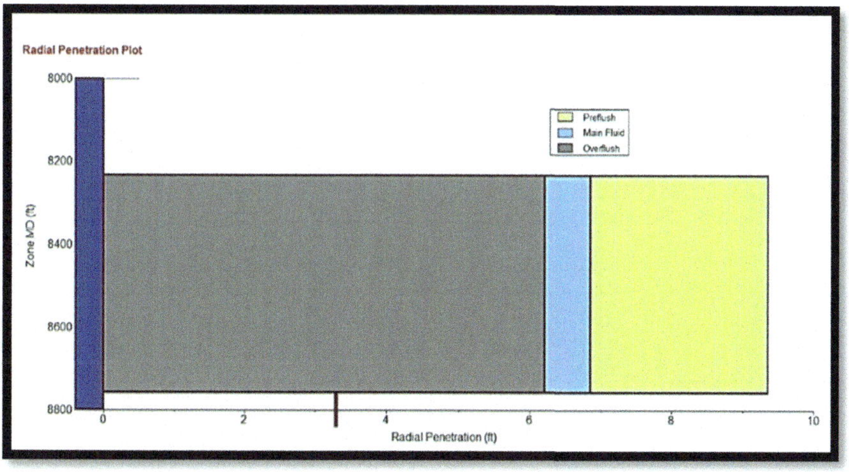

Fig. 4.16 Radius of penetration diagram for 7.5% HCl acid + 2.5% HAc

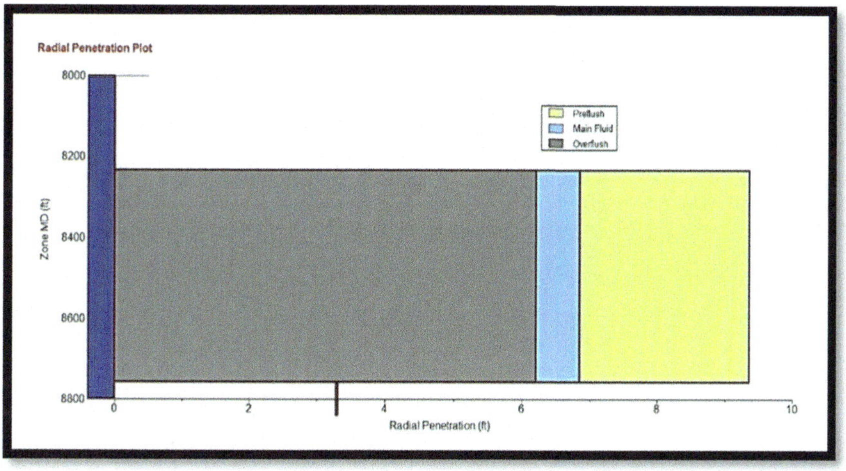

Fig. 4.17 Radius of penetration diagram for 15% HCl acid

4.5.2.5 Graphs of Pressure and Injection Rate Over Time

Figures 4.19, 4.20, 4.21 and 4.22 illustrate the variations in bottom well pressure, wellhead pressure, pressure break, and injection rate at each stage relative to the time spent in each stage for the following acid compositions: HCl 7.5%, HCl 7.5% + HAc 2.5%, HCl 15%, and HCl 15% + HAc 13%.

In matrix acidizing, the objective isn't to generate fractures, but rather to enhance permeability by dissolving the rock matrix. The injection pressure is determined based on the formation's fracture pressure and the pressure drop occurring along

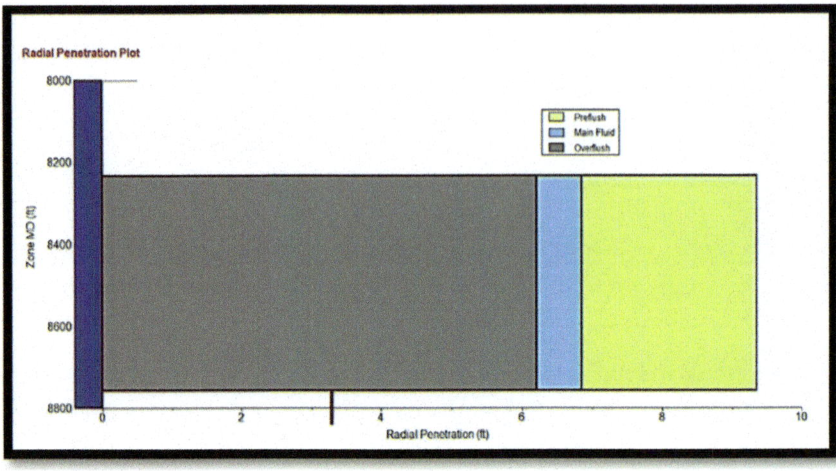

Fig. 4.18 Radius of penetration diagram for 15% HCl + 13% HAc acid

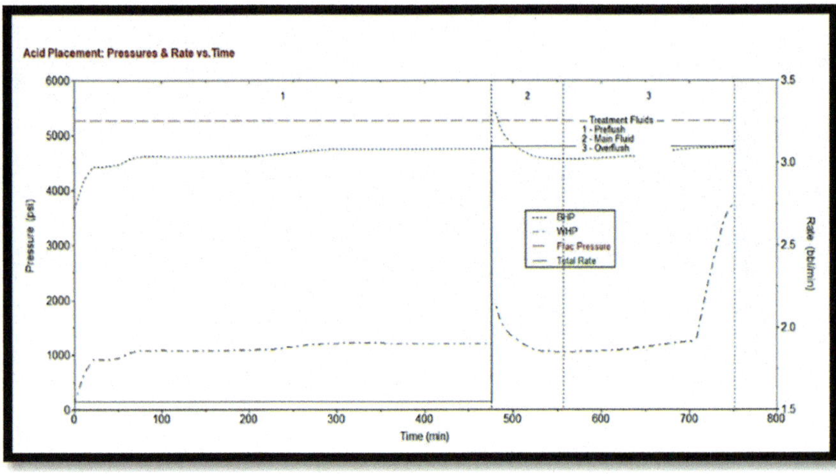

Fig. 4.19 Graph of pressure and injection rate in terms of time for 7.5% HCl acid

the well. This pressure represents the maximum limit that can be applied without causing formation damage, and it's essential to always maintain injection pressure below this threshold. The formation's fracture pressure also influences the maximum injection flow rate, which is the highest rate feasible. To achieve optimal acidizing, it's imperative to create high-permeability pathways, typically achieved by forming wormholes. Figures 4.19, 4.20, 4.21 and 4.22 depict formation fracture pressure, wellhead pressure, well bottom pressure, and injection flow rate over time. The formation's fracture pressure, typically around 5000 psi, can be determined based on the well's depth. Upon initiating injection, wellhead pressure should be adjusted

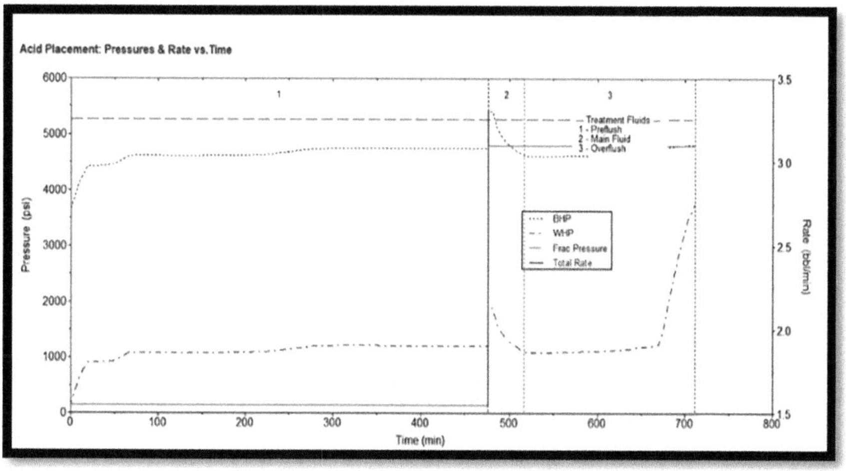

Fig. 4.20 Graph of pressure and injection rate in terms of time for HCl acid 7.5% + HAc 2.5%

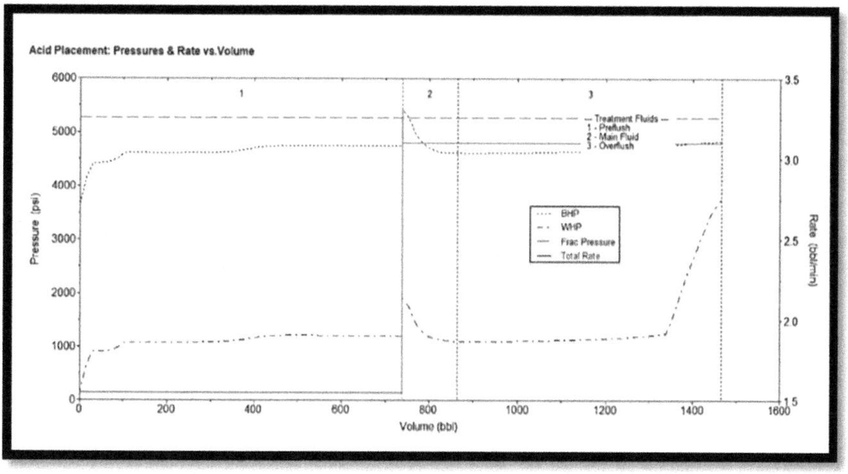

Fig. 4.21 Graph of pressure and injection rate in terms of time for 15% HCl acid

in a manner that, when combined with hydrostatic pressure, results in a bottomhole pressure lower than the formation's fracture pressure.

Example 4.1 The well is completely vertical and is producing carbonate and 100% calcite. The damage is caused by the well completion operation and the depth of the formation damage is 8 inches.

According to the input data of the software, the amount of skin is obtained in 3 stages.

The simulation is performed in 6 steps as follows:

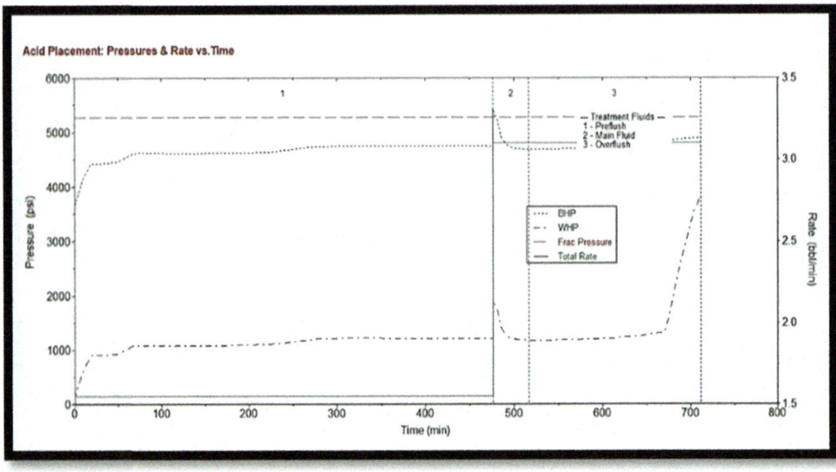

Fig. 4.22 Pressure graph and injection rate in terms of time for 15% HCl + 13% HAc acid

1-First Step: Administrative Data (Company, Well and Field Name, Location, etc).
This section shows a summary of the simulated well. (Complete the table using the example) (Fig. 4.23).

2-Second Step: Well completion Data (Perforations, Hole Survey, Tubular, etc.) (Fig. 4.24).

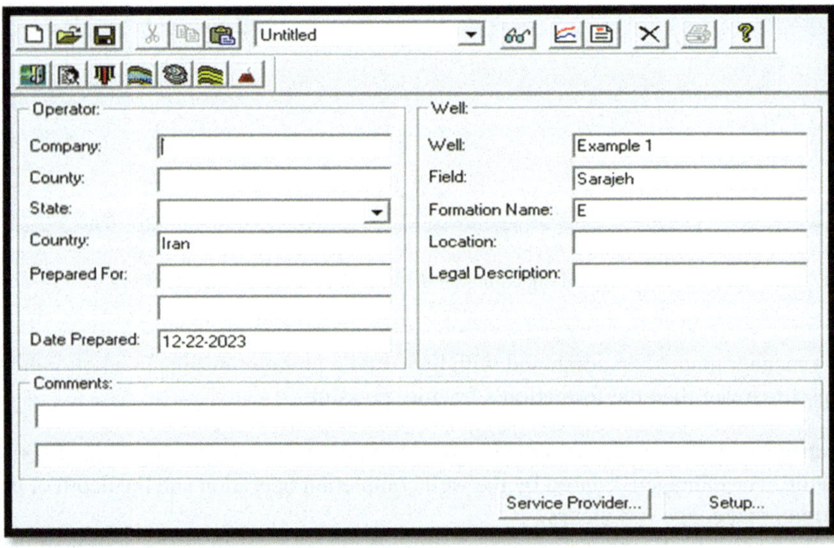

Fig. 4.23 Administrative data

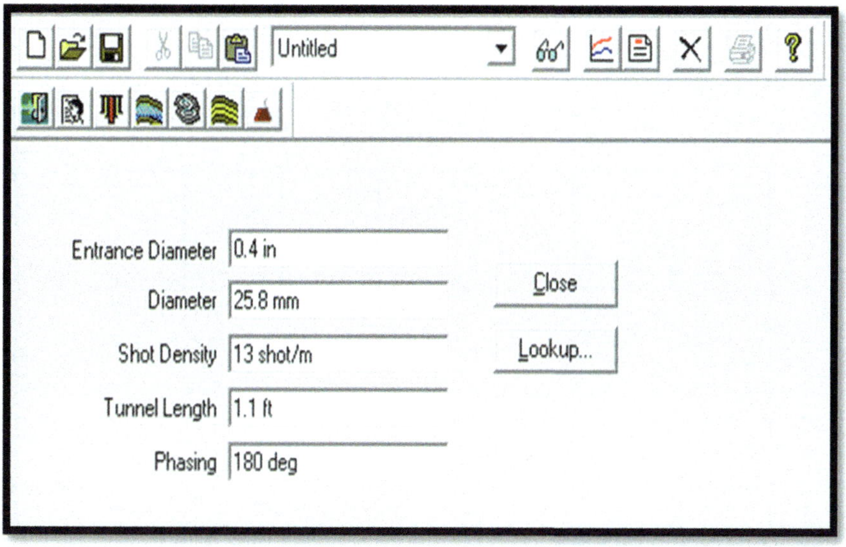

Fig. 4.24 Perforation data

Data related to this part: Injection is from annulus tubing. The type of well completion is completion by perforation. whose information is as follows.

The packer depth value, equal to 10687 ft, is entered after the perforation information is entered.

This section is used to enter casing and tubing information as follows. As shown in the Figs. 4.25 and 4.26.

3-Third Step: Reservoir Fluid Data (PVT Table or General Data Includes P_b, API Brine and Gas Composition).

Enter the data from this section into the software as shown in the Fig. 4.27.

4-Fourth Step: Reservoir Characteristic (P_r, Drainage Area, r_e, T, S_w, S_o, S_g).

The reservoir data is as follows in Figs. 4.28 and 4.29.

5-Fifth Step: Zone Data (Interval, Skin Factor, Porosity, Permeability, Fracture Gradient, Lithology, etc.).

The log data is as follows in Fig. 4.30.

The amount of lithology is shown in the figure. As the formation is carbonate, the amount of calcite is considered to be 100% (Fig. 4.31).

As said before. The damage in the well is due to the completion of the well and we choose the penetration damage of 8 inches. It is shown in the Fig. 4.32.

6-Sixth Step: Stimulation Fluid Data (Fluids Type, Additives).

At this stage, the information of the fluids to be used in acidizing should be selected.

The total amount of fluids available is shown in Table 4.4, which is as follows.

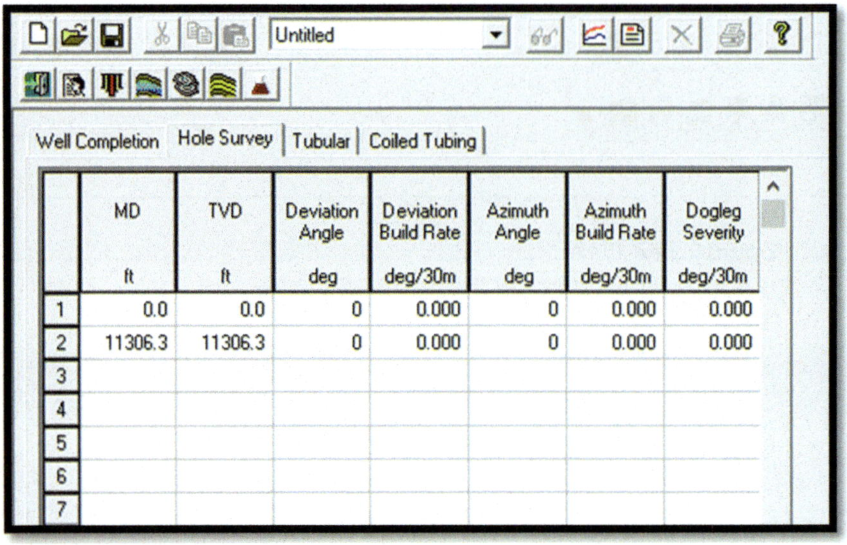

Fig. 4.25 Hole survey

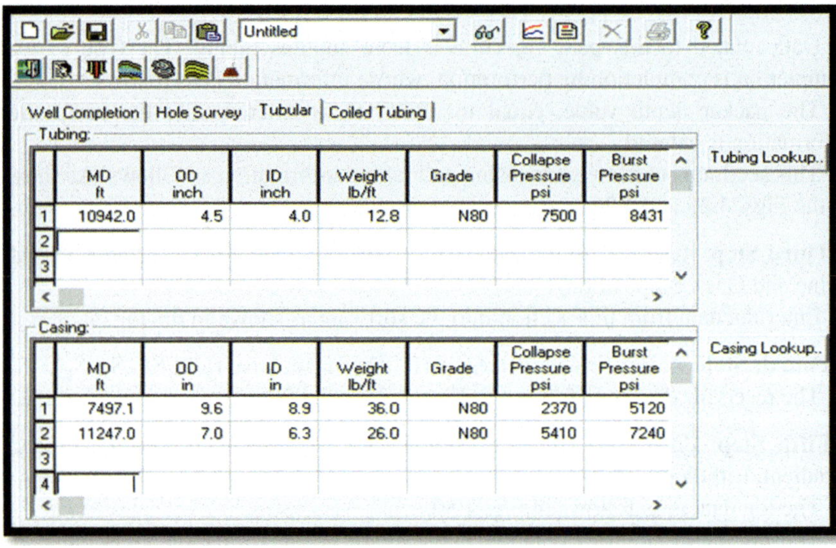

Fig. 4.26 Casing and tubing information

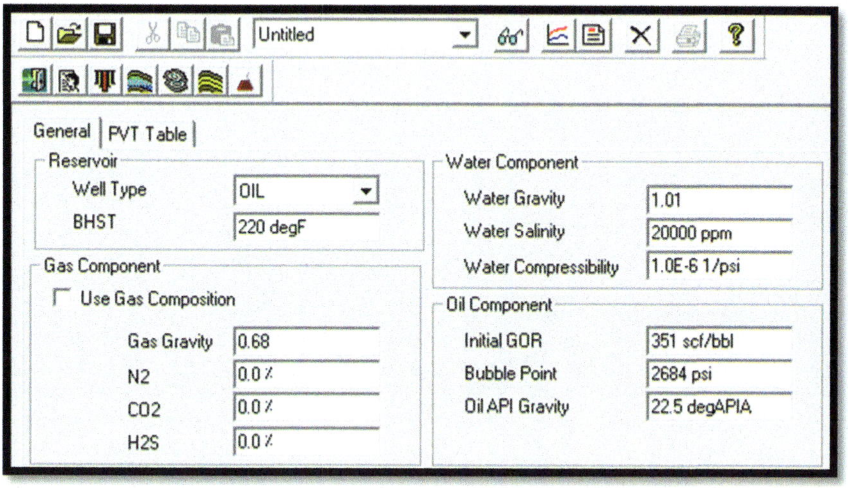

Fig. 4.27 Reservoir fluids section

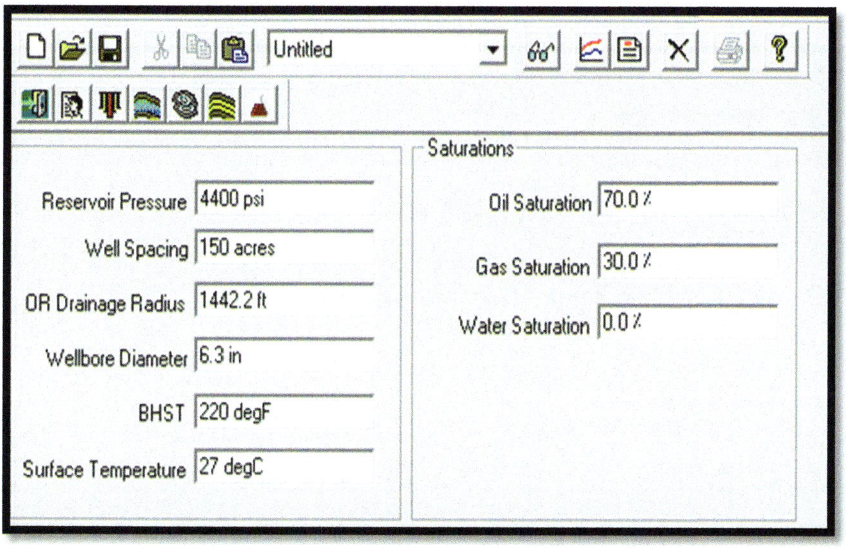

Fig. 4.28 Reservoir data Part 1

After selecting the Pumping Schedule Generator stage, new windows will open in which, in the first step, the viscosity and liquid gravity must be entered, which are 4.36 cp and 0.92 respectively which can be seen in the Fig. 4.33.

In the next part, the given skin (skin damage = 0.8) is entered to see the total skin value. It can be seen in the Fig. 4.34.

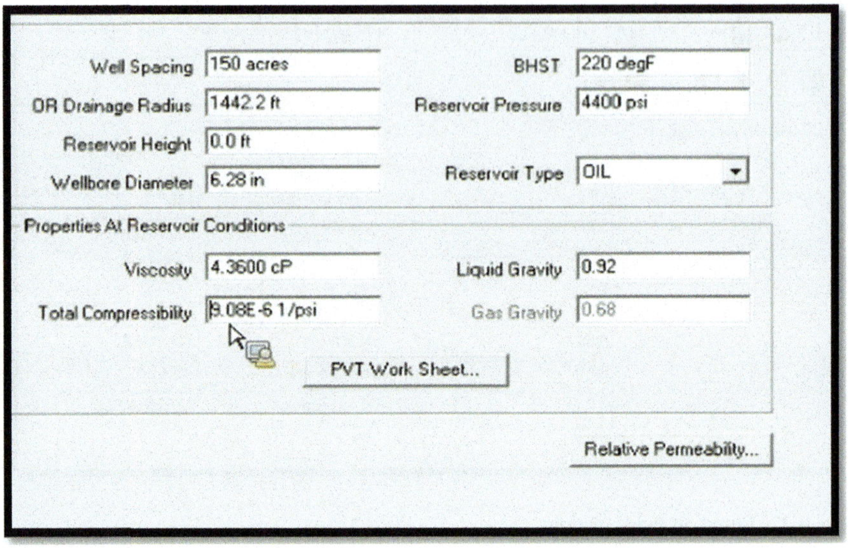

Fig. 4.29 Reservoir data Part 2

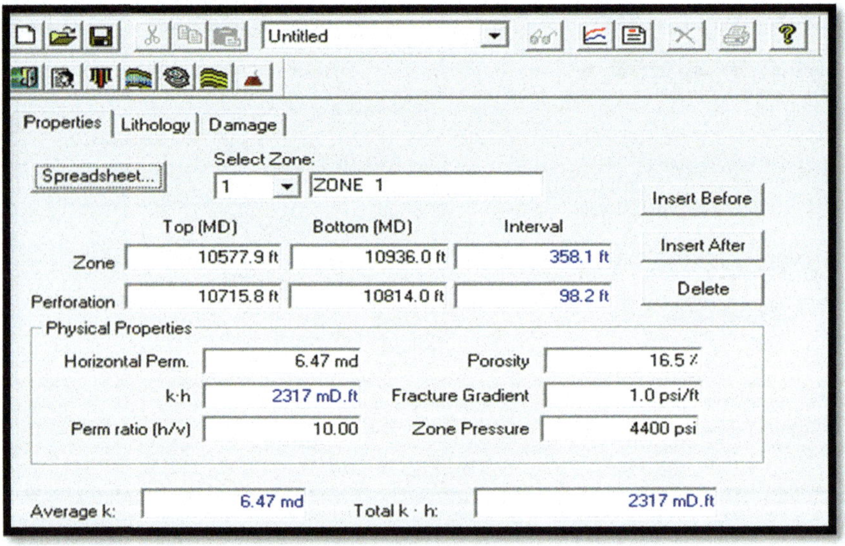

Fig. 4.30 Zone properties

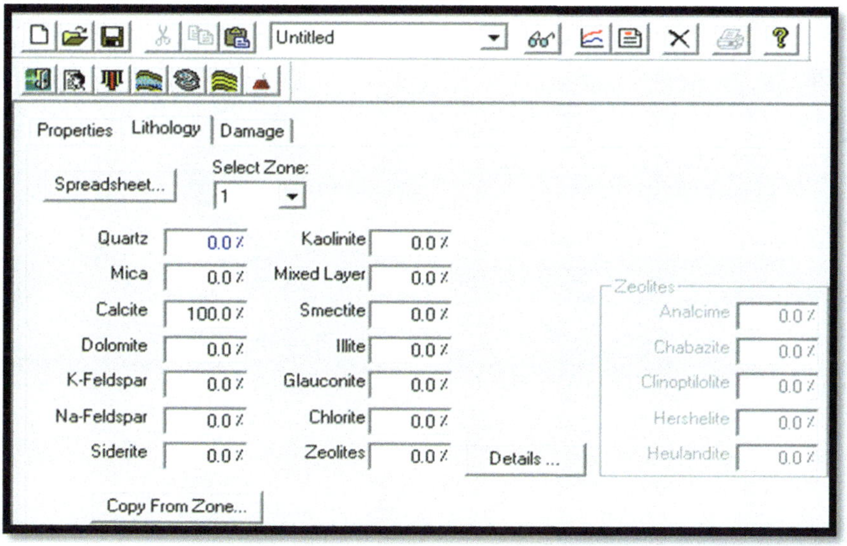

Fig. 4.31 Lithology properties

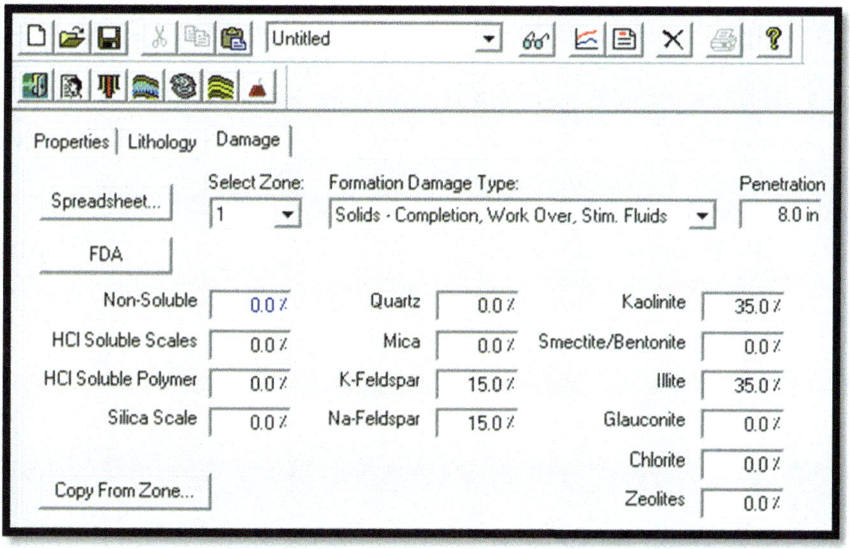

Fig. 4.32 Damage property

Table 4.4 Fluid information

Treatment Fluids	
Disel(gal)	Max 22800
HCl 15% (gal)	Max 8000
Emulsion acid (gal)	Max 20000

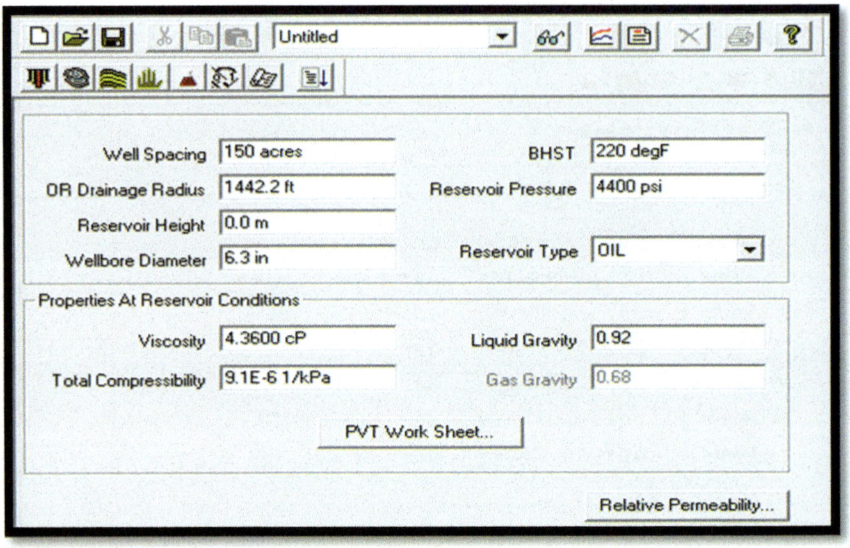

Fig. 4.33 Edit zone properties data

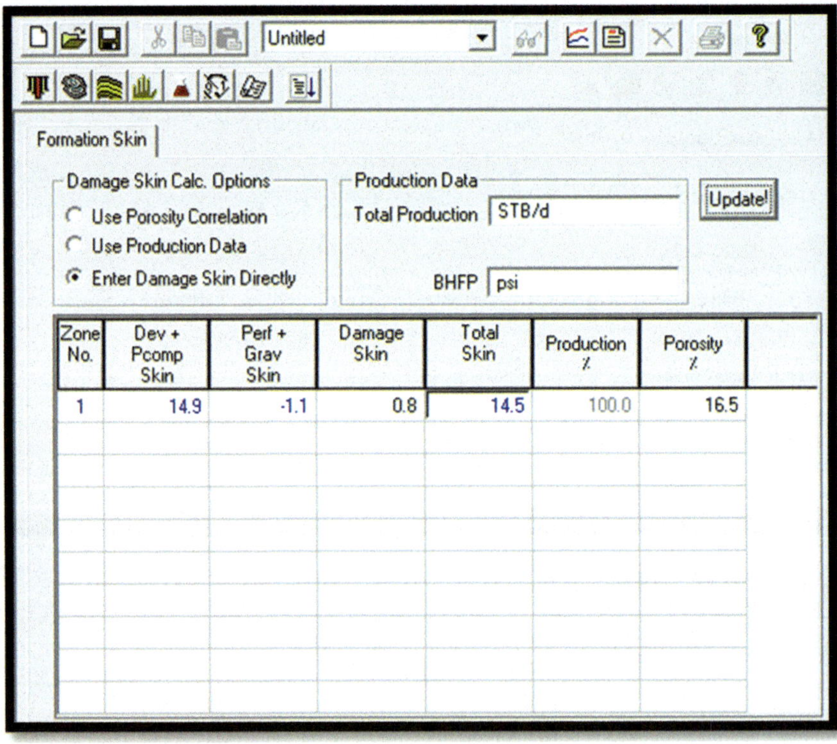

Fig. 4.34 Edit skin analysis data

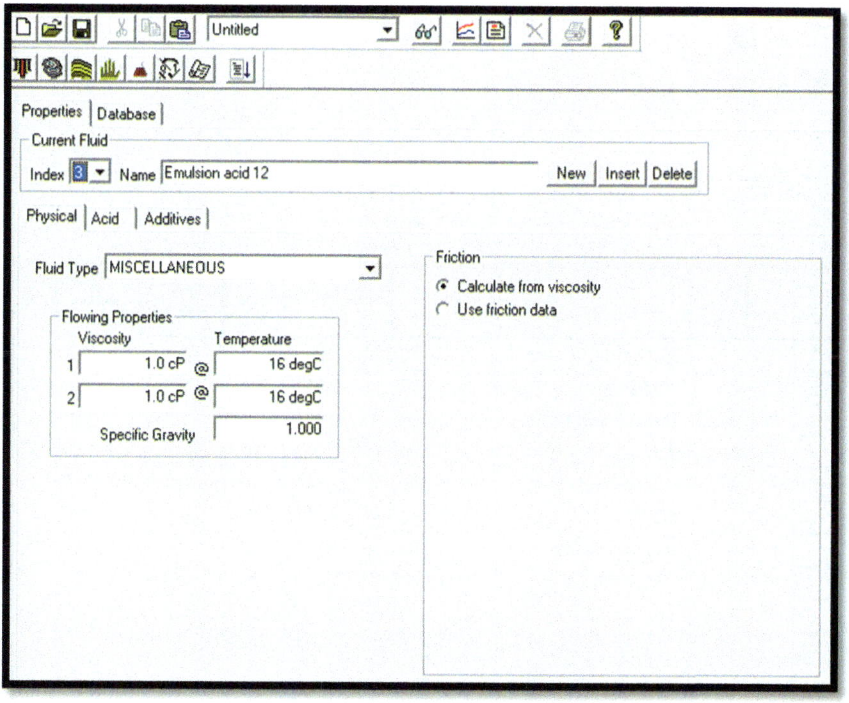

Fig. 4.35 Edit stimulations fluid data

In the next part, three types of fluids (Diesel, HCl 15% and Emulsion Acid) are entered into the data in three indexes. which is shown in the Fig. 4.35.

In the next step, the fluids selected in 8 steps are entered as (Pre-Flush, main-flush and over-flush). Injection flow rate and acid volume are suggested by the software (These two items can be changed manually with desired data) (Fig. 4.36).

In the final part, acid placement should be selected to run from all the data that has been entered, and then the final results that can be seen in the Fig. 4.37. Results like final damage, total pump time is visible.

And also some charts Figs. 4.38, 4.39, 4.40 are shown. It should be noted that different results and graphs can be prepared from this software and only a few of them are included in the figures.

4.6 Questions

Question 4.1 In a vertical oil and gas well, find the value of the skin coefficient in each of the following situations:

A: 2 injection stages with and without NH_4Cl

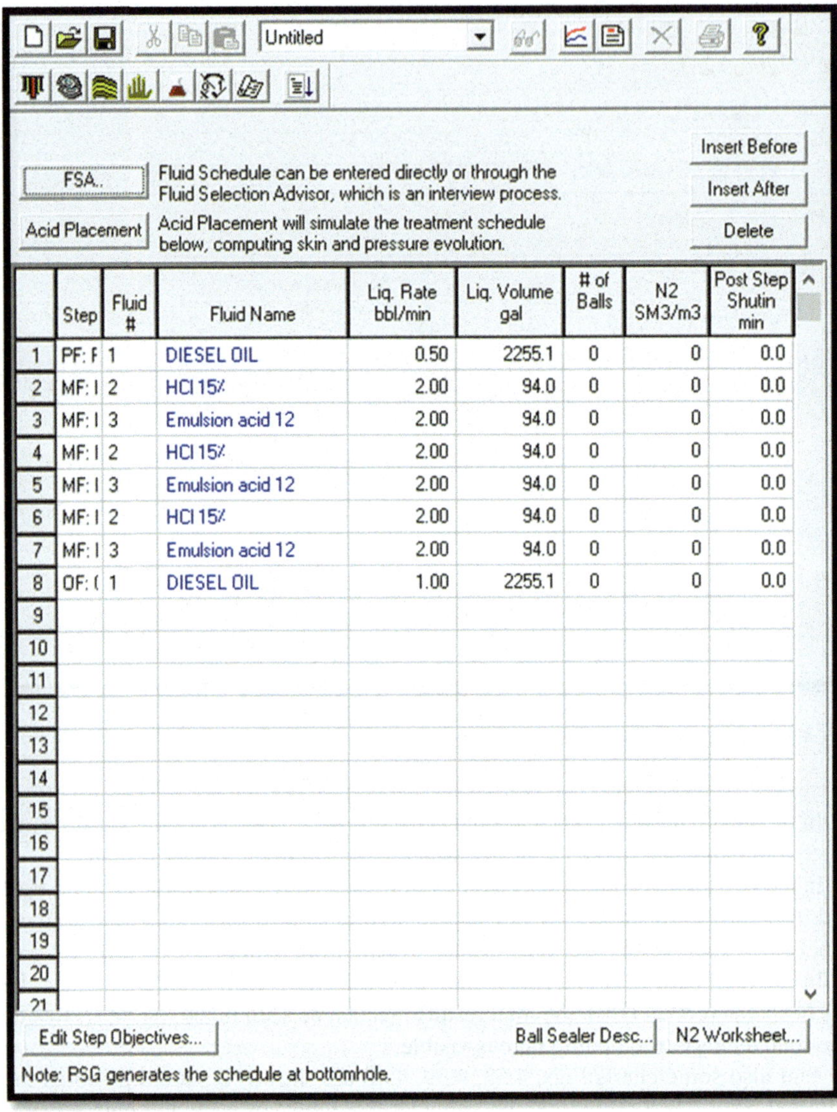

	Step	Fluid #	Fluid Name	Liq. Rate bbl/min	Liq. Volume gal	# of Balls	N2 SM3/m3	Post Step Shutin min
1	PF: F	1	DIESEL OIL	0.50	2255.1	0	0	0.0
2	MF: I	2	HCl 15%	2.00	94.0	0	0	0.0
3	MF: I	3	Emulsion acid 12	2.00	94.0	0	0	0.0
4	MF: I	2	HCl 15%	2.00	94.0	0	0	0.0
5	MF: I	3	Emulsion acid 12	2.00	94.0	0	0	0.0
6	MF: I	2	HCl 15%	2.00	94.0	0	0	0.0
7	MF: I	3	Emulsion acid 12	2.00	94.0	0	0	0.0
8	OF: (	1	DIESEL OIL	1.00	2255.1	0	0	0.0
9								
10								
11								
12								
13								
14								
15								
16								
17								
18								
19								
20								
21								

Fig. 4.36 View schedule data

B: 4 injection stages with and without NH_4Cl

The input data to the software are as follows.

The mineral content of the formation is 60% calcite and 40% dolomite. Also, the fluids used in the acidizing process are NH4Cl 3%, HCl 15%, diesel oil.

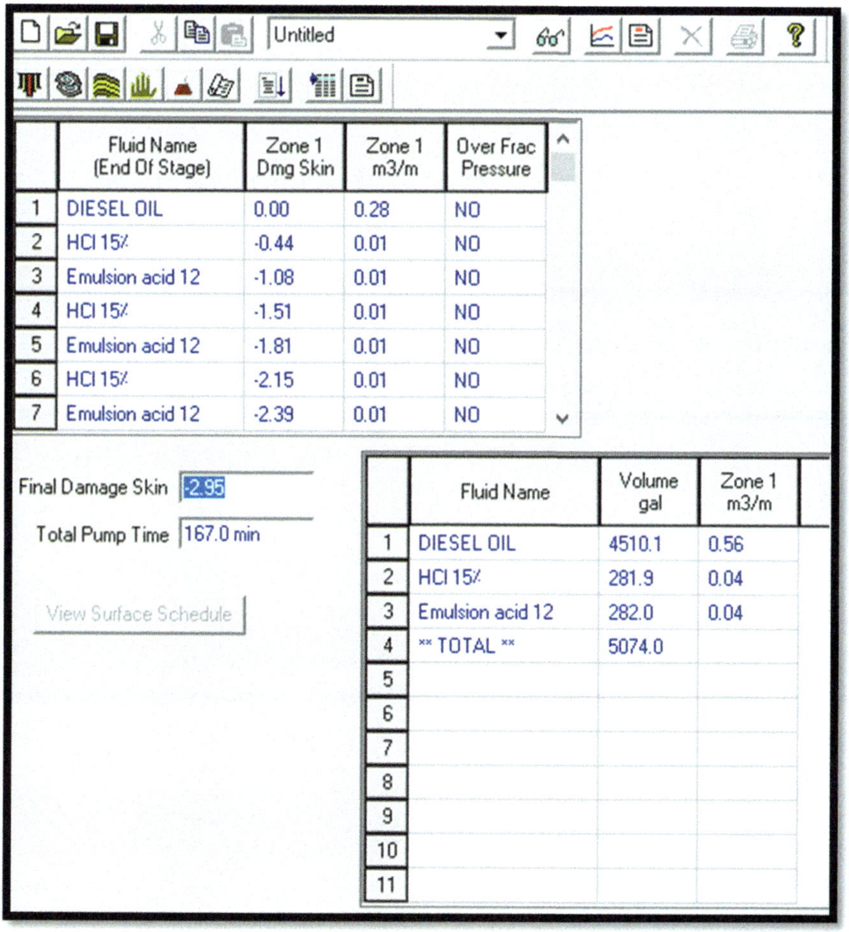

	Fluid Name (End Of Stage)	Zone 1 Dmg Skin	Zone 1 m3/m	Over Frac Pressure
1	DIESEL OIL	0.00	0.28	NO
2	HCl 15%	-0.44	0.01	NO
3	Emulsion acid 12	-1.08	0.01	NO
4	HCl 15%	-1.51	0.01	NO
5	Emulsion acid 12	-1.81	0.01	NO
6	HCl 15%	-2.15	0.01	NO
7	Emulsion acid 12	-2.39	0.01	NO

Final Damage Skin -2.95

Total Pump Time 167.0 min

View Surface Schedule

	Fluid Name	Volume gal	Zone 1 m3/m
1	DIESEL OIL	4510.1	0.56
2	HCl 15%	281.9	0.04
3	Emulsion acid 12	282.0	0.04
4	** TOTAL **	5074.0	
5			
6			
7			
8			
9			
10			
11			

Fig. 4.37 Information obtained after fluid injection

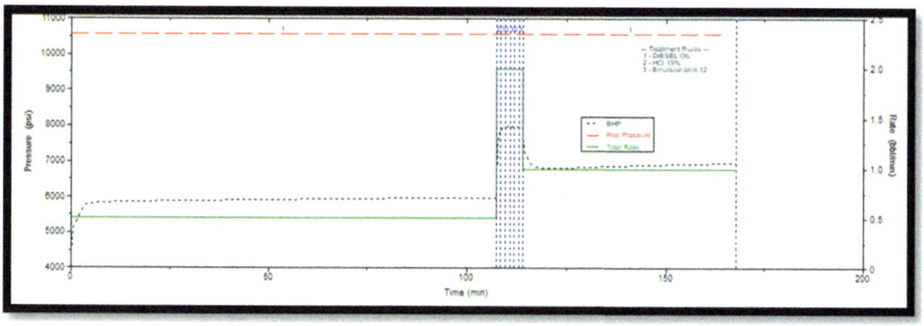

Fig. 4.38 Pressure versus time graph

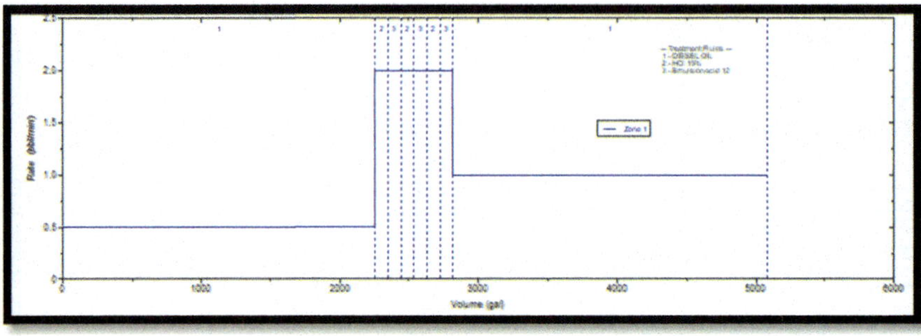

Fig. 4.39 Flow rate versus volume

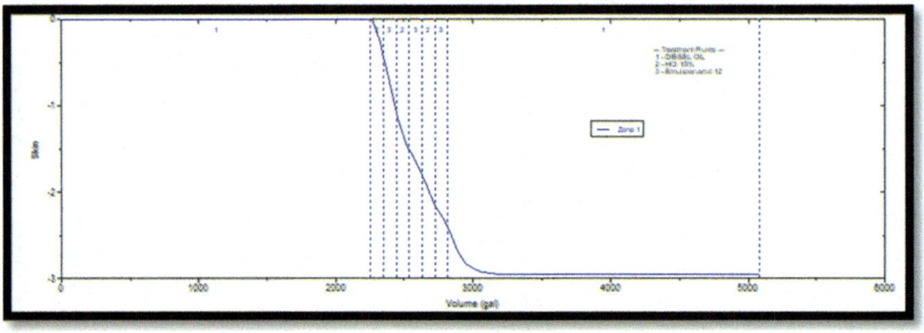

Fig. 4.40 Skin versus volume

	MD (m)	OD (in)	Grade
Casing	3550	7	N80
Tubing	3166	4.5	H40

MD = TVD = 3550

Perforation: 3190–3200 m

Entrance diameter	0.3 in
Diameter	8.4 mm
Shot density	13 shot/m
Tunnel length	0.05 ft
Phasing	120 deg

Reservoir properties	
Porosity	20%

(continued)

(continued)

Reservoir properties	
Permeability	36 md
Fracture gradient	0.69 psi/ft
Damage	Fine migration
Skin	0.4
Penetration	2.5 ft
Pressure	3000 psi
Temperature	160 F
Well spacing	330 acres
Wellbore diameter	6.4 in
Compressibility	1.7E-4 1/psi
Zone	3184–3201.5 m
Radius of well	3.2 in

Fluid properties	
Viscosity	0.92 cp
SG gas	0.64
SG oil	0.89
Pb	2935.5 psia
API	27
Rp	541 SCF/STB
Bo	1.2 bbl/STB
Sg	40%
So	60%

Chapter 5
Economic Evaluation and Estimation

5.1 Introduction

Economic evaluation involves the assessment of an activity based on its costs and consequences. In any rational decision-making process, it's crucial to weigh the pros and cons. The objective of economic evaluation is to provide an estimation of a project's costs and benefits over time. The outcome of this evaluation can lead to three possible scenarios: continuing the project as-is, continuing it with anticipated modifications, or canceling the project altogether. There are several reasons for conducting economic evaluations of projects. Projects are considered economically efficient when their benefits outweigh their costs, and the most efficient project is one that maximizes this difference. In the oil, gas, and petrochemical industries, numerous economic forecasts are essential before initiating a project. These forecasts encompass estimating project startup costs, assessing the project's economic viability and profitability, identifying cost reduction strategies, and addressing potential challenges. In the demanding realm of the oil and gas industry, the primary challenge often lies in optimizing the exploitation of these vast and valuable resources. By streamlining the operating costs of oil and gas reservoirs, it's possible to reduce these expenses to some extent, ultimately increasing profitability from these resources. Within the context of acidizing processes, economic evaluation and cost estimation hinge on various factors and parameters. These include the time required for acidizing and the volume of acid used. Furthermore, precise and up-to-date pricing information for necessary equipment, materials, labor, and global oil and gas prices is essential for performing accurate profit and loss calculations.

The prices of oil and gas are influenced by three primary groups of factors.

1. Supply and Demand for Each Period: Short-term price fluctuations are heavily influenced by the immediate supply and demand dynamics within the market.
2. Unforeseen Events and Accidents: Unexpected occurrences like wars, refinery fires, or oil tanker accidents can have a significant impact on prices by disrupting the normal supply chain.

© The Author(s), under exclusive license to Springer Nature Switzerland AG 2024

E. Khamehchi et al., *Applied Matrix Acidizing of Carbonate Reservoir*,
Petroleum Engineering, https://doi.org/10.1007/978-3-031-58281-3_5

3. Market Psychology: The psychological conditions and sentiments prevailing within the market can also play a role in influencing short-term prices.

Over the long term, price trends are primarily driven by the first factor (supply and demand) in conjunction with the policies and strategies of major producing and consuming countries. These countries can have a substantial impact on the overall market dynamics and pricing trends in the extended term.

5.2 Economic Calculations

5.2.1 Essential Costs for Acidizing

In an acidizing operation, acid is injected into the wellhead or through a coiled tubing. Before each injection, it's imperative to prepare the necessary equipment and materials. Fixed factors and parameters in each operation encompass the required acid volume, water volume, necessary additives volume, acid storage reservoirs, injection pump, and labor. Accurate knowledge of these values is essential for conducting economic calculations. The materials involved include various types of acids such as hydrochloric acid, acetic acid, methanol, diesel, and water (without solvents). The cost estimation of additives employed in acidizing should be precise. Additives like iron control agents, surface tension reducers, non-emulsifiers, corrosion inhibitors, acid corrosion inhibitors, Acid corrosion inhibitor intensifier aid, and mutual solvents are essential in acid treatment. For acidizing, various storage reservoirs and warehouses are necessary to transport materials in volumes ranging from 300 to 500 barrels. Additionally, specialized personnel are required to execute operations, with costs and salaries expressed on a per-day basis. The expenses associated with moving and transporting equipment, storage tanks, and essential materials should be calculated before commencing operations. If a camp and adequate personnel equipment are needed for the acidizing process, camp costs should be factored in for each day Fig. 5.1 illustrates the required equipment and tools in the acidizing process. The cost of all these items must be considered.

To conduct calculations and estimate the essential costs for acidizing operations, it's crucial to have information about the acidizing duration and the volume of acid required. Typically, a period of 5 days is allotted for acquiring equipment, materials, labor, transportation, and executing the acidizing process. The necessary acid volume is determined, and based on this acid volume, various factors are determined, including the number of workers, storage space for materials and additives, and the volume required for acid, diesel, and water. Additionally, this information helps in determining the number of necessary equipment and pumps.

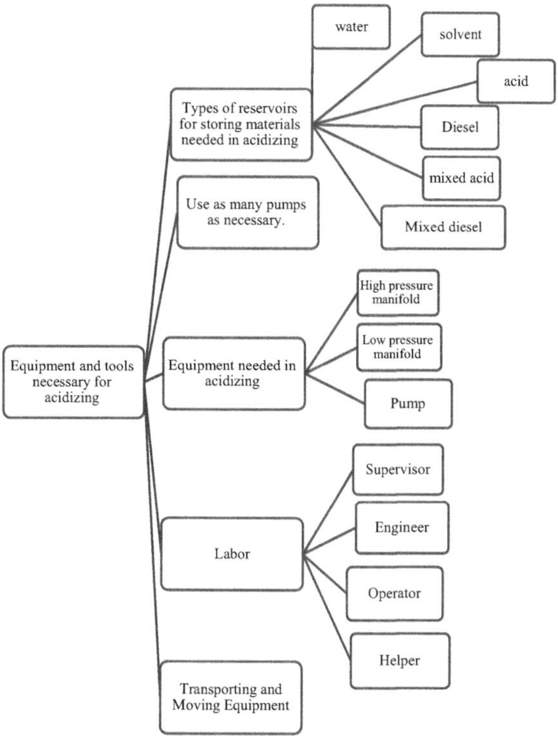

Fig. 5.1 A complete schematic of equipment and tools necessary in acidizing

5.2.2 *Flow Rate Changes After Acidizing*

In a production system, fluid travels from the reservoir into the well and is brought to the surface through the well, subsequently passing through two and three-phase separators to separate water, oil, and gas. When reservoir fluid enters the well, it must possess the requisite energy and capability to ascend to the surface. Occasionally, due to damage within the formation, the reservoir fluid experiences an additional pressure drop. While it does reach the surface, it is not optimally economically produced. This necessitates the utilization of well-stimulation methods to enhance production rates. The process involves injecting acid and repairing the damage, ultimately improving the reservoir's production performance.

To assess the increase in production following acidizing, the well's performance is assumed to remain constant. Reservoir performance graphs, varying in permeability, and a well performance graph are generated. By examining the point of intersection, changes in production flow rates can be observed. In this research, PROSPER software was employed to simulate production performance, encompassing reservoir flow, well flow dynamics, and production flow rates. Sensitivity analyses were conducted on initial permeability, permeability post-damage, and permeability post-acidizing. Reservoir and well performance graphs were plotted to visualize the shifts in production flow. Figure 5.2 illustrates the performance diagram of the reservoir and well under three conditions: initial permeability, permeability post-damage, and permeability post-acidizing.

As depicted in Fig. 5.2, initially, the permeability experiences a decrease due to the injection of condensate. Subsequently, following the flooding operation, permeability increases, leading to improved reservoir performance. This results in an intersection with the well performance graph at higher flow rates.

Table 5.1 provides information regarding the initial flow rate, the flow rate subsequent to formation damage, and the flow rate following well stimulation.

As indicated in Table 5.1, following acidizing, the production flow increases by 85% compared to the flow rate after formation damage and by 20% in comparison to the initial permeability.

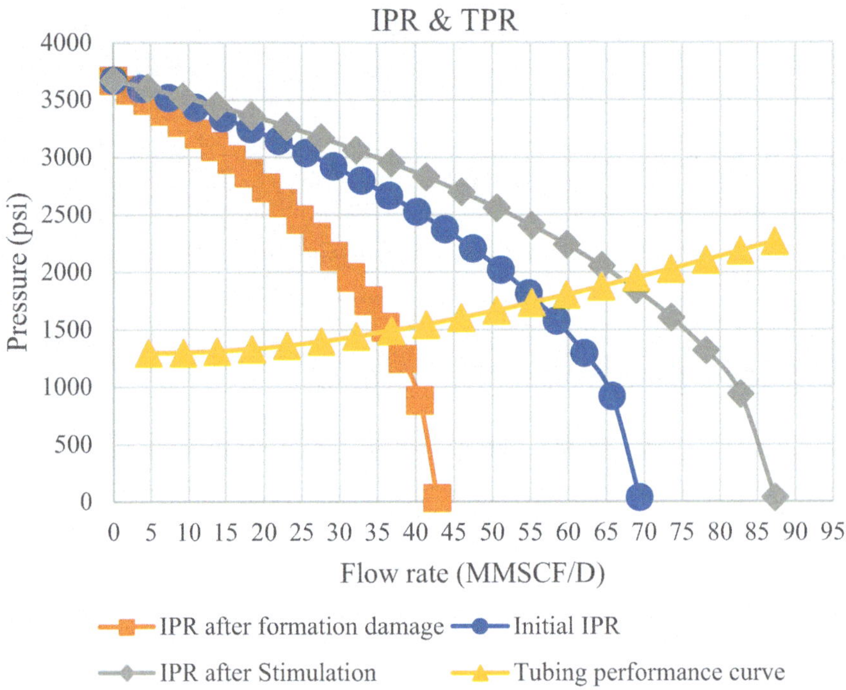

Fig. 5.2 Reservoir and well performance diagram

Table 5.1 Changes in production flow

Flow changes	Flow (MMSCF/D)
Initial flow	55
Flow after damage	36
Flow after stimulating the well	67

5.2.3 *Estimation of Costs and Profits Resulting from Acidizing*

To execute acidizing operations, an initial expenditure is required to prepare the necessary materials and equipment. However, following acidizing and the subsequent increase in production rates, this initial capital outlay is expected to be recouped over a certain period. To perform these calculations accurately, obtaining precise gas price data is imperative.

Example 5.1 A gas well is experiencing condensate blocking damage, resulting in a flow rate of 36 MMSCF/D. A strategic remedy involving the utilization of 15% hydrochloric acid (HCl) is employed to alleviate the damage. Post-treatment, the well's flow rate is augmented to 67 MMSCF/D. Calculate the return on investment of the acidizing operation based on the following information.

- Initial well flow rate: 36 MMSCF/D.
- Post-treatment well flow rate: 67 MMSCF/D.
- Consumption costs from Table 5.2.
- Fluid volumes and associated costs from Table 5.3.
- Consider the price of condensate at $8000 per ton.

Table 5.2 Worker, equipment, camp, mobilization and pump costs

Required components in acidizing	Prtice ($)
Equipment	500
Worker	1250
Pump	2000
Storage	4500
Camp	5000
Mobilization	20,000
Disel	24,000
Total	56,750

Table 5.3 Volume and cost of materials required for main acid, preflush and postflush

	Volume (gal)	Cost ($)
HCl 33%	6215	930
Additives	820	9585
Water	6635	630
Disel	8400	24,000
Total	22,070	35,145

Answer:

Cost of Acidizing

Cost of Components Required for Acidizing = 56,750 $

Cost of Materials Required for Main Acid, Preflush and Postflush = 35,145 $

Total Costs = 91,895 $

Income of Acidizing

$$Revenue\ 1\ MMSCF\ Condensate = \frac{43.42 \times 10^6\ \text{Btu}}{1\ \text{tonne of condensate}} \div \frac{1055\ \text{Btu}}{1\ \text{standard cubic foot of natural gas}} \times \frac{\$\ 8000}{1\ \text{tonne of condensate}} \times \frac{1\ \text{MMSCF}}{10^6\ \text{standard cubic feet}} = \$\ 328.76$$

Increase the Amount of Condensate = 67 MMSCF/D-36 MMSCF/D = 31 MMSCF/D

$$One day's income from acidizing operation = \frac{31\ \text{MMSCF}}{1\ \text{day}} \times \frac{328.76\$}{1\ \text{MMSCF}} = 10191.56\$$$

Return on Investment

Time (day)	Cost ($)	Income ($)	Cost and income difference ($)
1	91,895	10,191.56	−81,703.44
2	91,895	20,383.12	−71,511.88
3	91,895	30,574.68	−61,320.32
4	91,895	40,766.24	−51,128.73
5	91,895	50,957.8	−40,937.2
6	91,895	61,149.36	−30,745.64
7	91,895	71,340.92	−20,554.08
8	91,895	81,532.48	−10,362.52
9	91,895	91,724.04	−170.96
10	91,895	101,915.6	10,020.6

The cost of acidizing is recouped within 10 days due to increased condensate production.

5.3 Fold of Increase (FOI)

The Fold of Increase (FOI) is a measure used in the oil and gas industry to evaluate the effectiveness of an acid job, which is a well stimulation technique used to enhance production from reservoirs. The FOI is calculated by comparing the post-acid treatment production rate to the pre-treatment production rate. From a technical perspective, the FOI provides valuable insights into the success of an acid job. A higher FOI indicates a greater increase in production, suggesting that the acid treatment was effective in removing formation damage and improving reservoir permeability. This evaluation helps operators determine whether additional acid treatments are required or if alternative stimulation techniques should be considered.

Economically, the FOI plays a crucial role in assessing the financial viability of an acid job. The cost of performing an acid treatment can be significant, including expenses related to chemicals, equipment, and manpower. Therefore, it is essential to evaluate whether the potential increase in production justifies these costs. By comparing the FOI with other well stimulation techniques or alternative reservoir management strategies, operators can make informed decisions about allocating resources effectively. If the FOI is high and exceeds expectations, it may indicate that further acid jobs are warranted to maximize production from existing wells.

Conversely, if the FOI is low or does not meet expectations, operators may need to reassess their approach and consider alternative methods or technologies. Additionally, economic evaluation should also consider factors such as market conditions, oil prices, and overall field economics. It is important to assess whether the increased production resulting from an acid job will generate sufficient revenue to cover both immediate costs and long-term profitability.

Most fracturing jobs resulted in fold of increase, FOI, from 2 to 13. Economical evaluation showed that whole stimulation campaign recovered the investment during third month of execution (Izurieta et al., 2018).

In summary, both technical and economic evaluations are crucial when assessing the effectiveness of an acid job through FOI analysis. While technical evaluation focuses on understanding reservoir response and determining if further treatments are necessary, economic evaluation helps determine if the benefits outweigh the costs associated with performing these treatments.

In Fig. 5.3 a literature study was done through several sources to determine the optimal approach of acidizing design. Data preparation and acquisition are required for the well testing analysis, matrix acidizing design, and productivity index forecast after acidizing treatment. One of the challenges is a considerably limited number of available production history data. Well performance evaluations, such as well testing analysis, are used to identify the formation damage of the well. The calculation is based on pressure build-up data which includes calculating and analyzing permeability, skin factor, PI (in kg/hr/psi), flow efficiency, and radius of influence. Wellbore profile was calculated based on the analysis of fracture gradient, bottom-hole pressure fracture, maximum surface pressure, and maximum injection rate as input parameters for designing matrix acidizing in this well (Aurianto et al., 2022).

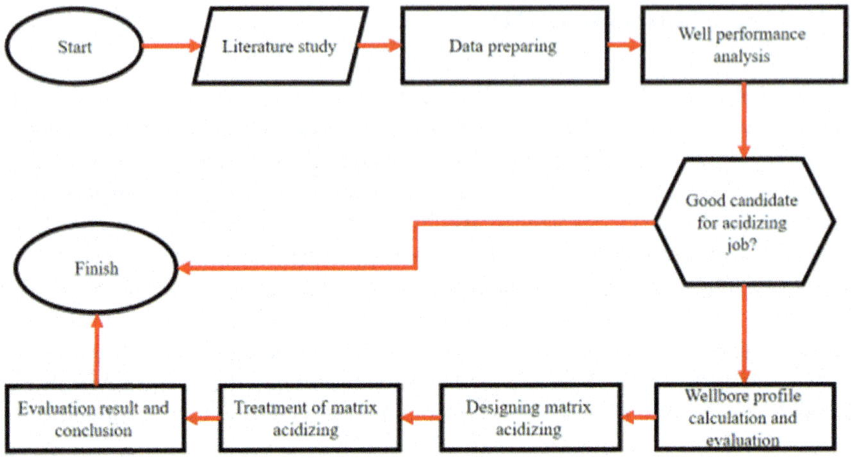

Fig. 5.3 Schematic flow diagram (Aurianto et al., 2022)

5.3.1 Maximize NPV Through Informed Decisions

All jobs were planned to reach ideal production case expectation, from design to execution. By setting the ideal cases as production goal, all team players delivered solutions to maximize the net present value (NPV).

The decision three shown in Fig. 5.4 highlights the possible scenarios during workover planning that helped to maximize NPV. This project used historical information and aimed to take new information to evaluate the productivity increase in each case, i.e. value information (Kamal et al., 2019). In some cases, the information showed high risk for project, at this point the asset team had possibility to make an informed decision as to whether the project should be dropped or workover should be redesigned according to the new data and if the risk was still favorable to production, the project should continue (Izurieta et al., 2018).

Execution

The executed workover program included up to these workover rigs with simultaneous operations on different project pads. The program can be summarized as follows.

1. Prepare location, the artificial lift, facilities and drilling rig according to the workover plan.
2. Pull out completion string, this operation could include fishing operations.
3. To isolate lower reservoirs to avoid crossflow communication through BHA.
4. Perform acid fracturing.
5. To evaluate the well with no rig while other fracture jobs start.
6. To close the well for pressure build up.
7. To complete the well with ESP.

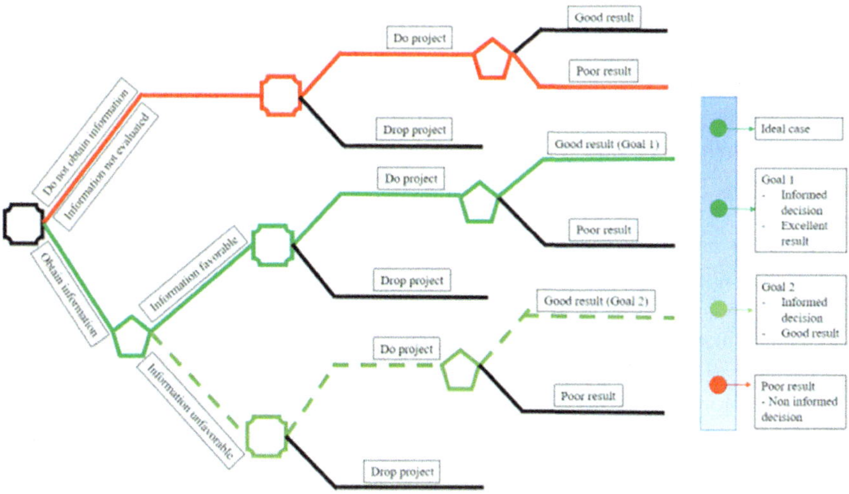

Fig. 5.4 NPV maximization though informed decisions and decision tree (Kamal et al., 2008)

Acidizing of carbonate reservoirs is a common technique used in the oil and gas industry to enhance production. However, it comes with certain issues and risks that need to be addressed. Here are some steps to address these issues, including risk assessments and post-stimulation evaluations.

1. Risk Assessment Process:

   a. a. Identify potential risks: Begin by identifying the potential risks associated with acidizing carbonate reservoirs. These may include formation damage, fluid loss, wellbore instability, environmental impact, etc.
   b. Quantify risks: Assess the likelihood and severity of each identified risk. This can be done through expert opinions, historical data analysis, and modeling techniques.
   c. Develop risk mitigation strategies: Once risks are identified and quantified, develop strategies to mitigate or minimize them. This may involve implementing best practices, using appropriate chemicals and additives, optimizing operational parameters, etc.
   d. Implement risk management plan: Incorporate the risk mitigation strategies into a comprehensive risk management plan that outlines responsibilities, procedures, and monitoring mechanisms.

2. Post-Stimulation Technical Evaluation:

   a. Data collection: Collect relevant data during and after the acidizing process, including pre- and post-stimulation well logs, pressure measurements, fluid samples, etc.
   b. Analysis of results: Analyze the collected data to evaluate the effectiveness of acidizing in terms of production enhancement and reservoir stimulation.

c. Identify limitations or issues: Identify any limitations or issues encountered during or after acidizing that may have affected the desired outcomes.
d. Recommendations for improvement: Based on the analysis and identified limitations/issues, provide recommendations for future improvements in acidizing techniques or operational practices.

3. Post-Stimulation Economic Evaluation:

a. Cost analysis: Evaluate the costs associated with acidizing operations including chemicals, equipment rental/maintenance, manpower requirements, etc.
b. Production analysis: Assess the impact of acidizing on production rates and overall recovery from the carbonate reservoirs.
c. Economic assessment: Calculate return on investment (ROI) by comparing increased production/recovery against costs incurred during acidizing operations.
d. Decision-making support: Provide insights into whether further acidizing operations are economically viable or if alternative methods should be considered.

4. Risk Assessment Tools:

a. Software applications: Utilize specialized software tools that can assist in conducting quantitative risk assessments by incorporating various parameters such as wellbore conditions, fluid properties, rock characteristics, etc.
b. Industry guidelines and standards: Refer to industry guidelines provided by organizations like API (American Petroleum Institute) or SPE (Society of Petroleum Engineers) that offer frameworks for conducting risk assessments specific to acidizing carbonate reservoirs.

By following these steps and utilizing appropriate tools and resources for risk assessment processes as well as post-stimulation technical and economic evaluations, operators can effectively address the issues associated with acidizing carbonate reservoirs while ensuring safe operations and maximizing production potential while minimizing costs.

References

Aurianto, M. P. D., Maulana, D. T., & Chandra, S. (2022). Evaluation of successful matrix acidizing method in a geothermal well with comparative sensitivity of acid fluid models, volume, and concentration: A case study on well "X". In *IOP Conference Series: Earth and Environmental Science*.

Izurieta, A., Cuenca, D., Pozo, L., Padilla, W., & Bustos, J. (2018). Economical mature field revitalization in low oil prices environment: A ten thousand incremental oil production case story in the drago complex field, Ecuador. In *SPE Trinidad and Tobago Section Energy Resources Conference?*

Kamal, A., Al-Ghamdi, S. G., & Koc, M. (2019). Revaluing the costs and benefits of energy efficiency: A systematic review. *Energy Research & Social Science, 54*, 68–84 August.

Kamal, M. M., Abbaszadeh, M., & Cinco-Ley, H. (2008). *Transient well testing*.